Julia Programming for Physics Applications

R. Gökhan Türeci · Hamdi Dağıstanlı ·
İlkay Türk Çakır

Julia Programming for Physics Applications

Springer

R. Gökhan Türeci
Institute of Accelerator Technologies
Ankara University
Gölbaşı, Ankara, Türkiye

İlkay Türk Çakır
Department of Accelerator Technologies,
Institute of Accelerator Technologies
Ankara University
Gölbaşı, Ankara, Türkiye

Hamdi Dağıstanlı
Radikal Konak Anatolian High School
Eşrefpaşa, İzmir, Türkiye

ISBN 978-3-031-84715-8 ISBN 978-3-031-84716-5 (eBook)
https://doi.org/10.1007/978-3-031-84716-5

This Springer imprint is published by the registered company Springer Nature Switzerland AG
The registered company address is: Gewerbestrasse 11, 6330 Cham, Switzerland

For Cemre, Elif and Karadut

—R. Gökhan Türeci

In memories of Günay Dağıstanlı and İlhami Sakarya

—Hamdi Dağıstanlı

For Eren

—İlkay Türk Çakır

Preface

Computer programming languages are used to express machine language codes consisting of 0's and 1's with various symbols and grammatical structures in microprocessor-driven systems. The grammatical structure created by the programmer with these symbols is different from the machine language and therefore must be converted into a form that the machine (microprocessor) understands. This conversion process is called compilation.

Computer programming languages are divided into three main groups according to their coding levels: low level, medium level and high level. In low-level software languages, machine language is expressed in symbols and is called Assembly language. The codes written in this language are long and turn into a shorter code whole after compilation. Assembly language is an efficient language. However, since it is not similar to human language structure, it is difficult both to learn this language and to code in this language. For this reason, intermediate software languages have been developed. Intermediate-level software languages are closer to human language structure and contain the features of low-level language. Therefore, it is close to both human language and machine language. Although the codes converted into machine language by compiling mid-level languages are not as effective as low-level languages, their performance is still good. Examples of intermediate-level languages are C or C++. High-level software languages are close to human language. Therefore, the learning and coding process is very short. These software languages have many software libraries. Thanks to these libraries, the software developer can benefit from these libraries without using long codes. Moreover, effective codes can be created by using the features in different libraries in different combinations according to the purpose. However, in high-level software languages, after the software is compiled, the number of machine language codes that arise due to the libraries used is high and therefore the performance of high-level software languages is low. However, since the processing power of microprocessors is high today, the processing power of the microprocessor tolerates the low performance of the software. Examples of high-level software languages are C#, Python, Julia and Java.

Software languages are also divided into three in terms of their operation; compliable, timely compiled, and interpreted languages. Examples of compiled languages are C, C++ and Pascal, and examples of timely compiled languages are C# and Java. In interpreted software languages, the code written by the developer is read line by line and executed by interpretation. Therefore, this language group needs an interpreter. One of the best examples of interpreted languages is Julia.

Julia software language is a high-level language in terms of coding type and interpreted language structure as a working principle. Since it has a high-level language structure, it is used effectively in artificial intelligence studies. However, another extremely important feature of the Julia language is that this language is an open-source language. Many software developers from all over the world make libraries available to users free of charge on platforms such as GitHub. Julia's characteristics can be summarised in general terms as speed, rich ecosystem, easy to use, parallelism, concurrency and cross platform compatibility. One of the most important features of Julia is its just-in-time (JIT) compilation. This allows Julia to dynamically compile code and optimise it for runtime performance. In addition to being a high-level language structure, being open source has made Julia a very popular language.

Although Julia is a language for general use and numerical computing, it is also well suited for data science and machine learning applications. Some of the main features of Julia are: dynamic type system, multiple references or multiple methods, integrated package manager, distributed and parallel computing, process management through the shell, support for user-defined types, simultaneous input and output processing, logging and performance analysis tools. Julia's dynamic type system allows the type of a value to change, and multiple dispatch or multiple methods can be used to call different methods at runtime. These features of Julia provide a great advantage for its use in data science and machine learning. In addition, Julia's features such as integrated package manager, distributed and parallel computing capabilities, logging and performance analysis tools make Julia preferable for data scientists and machine learning engineers.

To make academic calculations, there are software such as Mathematica, MATLAB, Maple, which are highly professional and work effectively, and which cost a high price to obtain with individual means. Mathematica, being a knowledge-based software, provides the opportunity to perform calculations without requiring much numerical analysis knowledge and is the most advanced software in its field. Similar calculations can be done in Julia language, which is open source, and using Julia libraries, which are also open source. However, the user must be familiar with the Julia language and moreover, the user must have a good command of numerical analysis.

In addition to being open source, Julia is a high-level language, which makes it possible to prepare highly efficient programmes. The best example of this is the considerable use of Julia language in artificial intelligence and machine learning studies.

In the introduction section of the book, the installation of Julia environment. In the next three chapters, general features of the Julia language are mentioned. In the fourth chapter, graphics specially prepared for numerical analysis are mentioned. Thus, it is tried to introduce plot modules and plot functions that can be used in physics and engineering calculations without writing extra code. In the sixth chapter, data frames are mentioned. Finally, in the next nine chapters, the features described in the first six chapters of the book are applied to physics (high energy physics, nuclear physics, solid state physics, mechanics, electromagnetism, astrophysics, quantum physics, waves) and statistics.

We believe that the book is useful.

Yahşihan, Türkiye — R. Gökhan Türeci
Eşrefpaşa, Türkiye — Hamdi Dağistanli
Gölbaşı, Türkiye — İlkay Türk Çakir

Competing Interests The authors have no competing interests to declare that are relevant to the content of this manuscript.

Contents

Part I
Julia Programming Properties

Chapter 1
Introduction

1.1 Why is Julia?

Julia is a high-level, general-purpose, and dynamic programming language with great features for data science, scientific computing, machine learning and artificial intelligence applications. Julia was designed from the ground up to be efficient and easy to use, and it has a rich set of libraries and tools for data analysis and visualisation.

Julia's development took a fresh approach to technical computing, combining the ease of use of Python and R with the performance of C and Fortran. Julia was designed from the ground up to address the needs of high-performance numerical computing, and its speed is on par with C++ and Fortran. In addition, the Julia code provides a convenient syntax that makes it easy to express mathematical ideas [1].

Julia is also open-source, so you can get started right away. And because Julia is designed for high-performance computing, it's a great choice for large-scale data analytics projects. So if you're looking for a powerful, easy-to-use, and open-source language, Julia is a great option.

One of the main reasons to use Julia is its speed. Julia's compiler technology allows it to achieve performance comparable to low-level languages like C or Fortran, while its high-level syntax makes it accessible to a wide range of users. This unique combination of speed and ease of use makes Julia an ideal choice for a variety of data science tasks [2, 3].

Julia's characteristics can be summarised in general terms as follows.

1. ***Speed***
   One of the most important advantages of using Julia is its speed. Julia is designed with performance efficiency in mind and can deliver results as fast as C and Fortran for a variety of applications. It has a sophisticated just-in-time (JIT) compiler that translates high-level code into machine code and optimises it for

R. G. Türeci et al., *Julia Programming for Physics Applications*,
https://doi.org/10.1007/978-3-031-84716-5_1

the system in use. It is therefore considered an ideal platform for large-scale data analysis, numerical computing and machine learning tasks.

2. ***Ease of Use***
   Julia's syntax is easy to learn and easy to read. In addition, it stands out as a high-level language that supports object-oriented, functional and meta-programming. Its similarity to Python makes Julia even easier to learn.
3. ***Rich Ecosystem***
   Julia meets a wide range of scientific computing and machine learning needs with a rich ecosystem of packages and libraries. It also includes a Package Manager, which makes it easy to install and manage packages, ensuring that you can always access the latest tools and libraries.
4. ***Interoperability***
   Julia has excellent interoperability with other languages and tools, making it a great choice for integration into existing data science workflows. You can easily call Python, C, and Fortran code from Julia and vice versa, allowing you to take advantage of existing libraries and tools without having to reinvent the wheel.

   Furthermore, Julia also has a Python and R package that enables you to use their libraries and tools directly in Julia.
5. ***Parallelism, Concurrency and Cross Platform Compatibility***
   Julia is designed for parallelism, concurrency and cross-platform compatibility, making it an ideal choice for modern data science workloads. Julia can run on all major operating systems, including Windows, macOS, and Linux.

   This means that you can write code once and run it on any platform, making it easy to collaborate and share your applications across a variety of environments. This can lead to significant performance improvements for many data science tasks.
6. ***Just-in-Time Compilation and Dynamic Computing***
   One of the most important features of Julia is its just-in-time (JIT) compilation. This allows Julia to dynamically compile code and optimize it for runtime performance.

   Therefore, these features provide the performance of a low-level language like C with the ease of use of a high-level language.
7. ***Open Source***
   Julia is an open-source language, which means that it is free for anyone to use, modify and distribute. This leads to a community of developers and users who support the use and development of the programme and ensure that it is constantly evolving.
8. ***Data Science and Machine Learning Features***
   Julia has the best libraries and packages for data science and machine learning tasks. Having libraries and packages to support its superior performance makes Julia ideal for data science and machine learning.
9. ***Graphical User Interface***
   Julia has several graphical user interfaces (GUIs) that you can use to create and run your programmes. One of the most important GUIs used in Julia is Julia

Studio, which provides a number of superior features such as code debugging, syntax highlighting and package management [2–4].

Although Julia is a language for general use and numerical computing, it is also well suited for data science and machine learning applications. Some of the main features of Julia are: dynamic type system, multiple references or multiple methods, Lisp-like programming language, integrated package manager, distributed and parallel computing, process management through the shell, support for user-defined types, simultaneous input and output processing, logging and performance analysis tools. Julia's dynamic type system allows the type of a value to change and multiple dispatch or multiple methods can be used to call different methods at runtime. These features of Julia provide a great advantage for its use in data science and machine learning. In addition, Julia's features such as integrated package manager, distributed and parallel computing capabilities, logging and performance analysis tools make Julia preferable for data scientists and machine learning engineers [5].

1.2 History of Julia

The Julia language was introduced in 2012 by a team of developers led by Jeff Bezanson, Stefan Karpinski, Viral B. Shah and Alan Edelman and has since been used in various fields, especially in machine learning, data science and scientific computing. There is no particular reason why the language is named Julia, the aim is to make it versatile and capable of performing different tasks. The design of the language started in 2009 and its goal was to create an open source, high-performance language suitable for scientific and numerical computing. The design philosophy of the language emphasises programmer productivity and code readability. In addition, Julia supports functional programming and metaprogramming in addition to traditional imperative and object-oriented programming.

The Julia programming language has a relative history. The important developments and the features gained are summarised below.

The Julia programming language dates back to 2009 and was developed by Jeff Bezanson, Stefan Karpinski, Viral B. Shah and Alan Edelman. The language aimed to combine the performance of low-level languages with the simplicity and flexibility of high-level languages. In 2012, the first public version was released and attracted attention with its performance and ease of use, especially in the field of numerical and scientific computing. Julia has been adopted by a wide community with its fast-working ability and big data processing features. Version 1.0, released in 2018, guaranteed that the code written to ensure stability and compatibility would remain functional in future versions. Since version 1.0, Julia has been continuously evolving and has been enriched with performance improvements, new features, and extended libraries. Use of the language has increased in academia and the community has grown.

Julia has been under continuous development since version 1.0. During this development process, performance improvements, new features and extended libraries have been added. The Julia community has also grown, and the language has become more widely used in academia, industry and various scientific fields.

Julia's key features include multiple references. This feature has become part of the language's appeal. Julia also differs from other high-level languages in terms of performance and speed. Its ability to compile Just-In-Time (JIT) and generate efficient machine code has made it the language of choice for performance-critical applications. The Julia community is actively working to create a rich package and library ecosystem. Julia also has the ability to interface with other languages (such as Python, R and C), making it a versatile tool in a variety of computational workflows [5].

1.3 The Environments of Julia

In Julia, the term 'environments' refers to different aspects of the language and its ecosystem. These environments include execution environments and package management systems. The main environments in Julia are the following:

1.3.1 Execution Environments

- REPL (Read-Review-Print Loop): Julia REPL is an interactive environment where you can execute Julia commands line by line. You can get immediate feedback, and it is useful for testing code snippets, performing quick calculations, and exploring Julia's features.
- Jupyter Notebooks: Julia can be used within Jupyter notebooks, where executable code, visualisations and narrative text are combined. This setup is popular for exploratory data analysis and sharing of interactive documentation.
- Integrated Development Environments (IDEs): Julia can be used in a variety of IDEs, such as Visual Studio Code, Atom, and JuliaPro, which offer features such as code completion, debugging, and project management.
- Command Line Interface (CLI): You can run Julia scripts and commands directly from the command line. This is useful for automating tasks and integrating Julia into larger workflows.
- IJupyter (also known as IJulia): IJulia, which allows you to run Jupyter notebooks with a Julia kernel (https://julialang.org/downloads/) [6]. After installing Julia, open the Julia REPL (Read-Eval-Print Loop), which is a command-line interface for running Julia code. On your system, this might be done by simply launching the Julia application, or by typing Julia in a terminal (Linux/macOS) or command prompt (Windows).

1.3.2 Package Management Environments

Pkg, Julia's built-in package manager, handles package installation and dependency resolution. Each project has its own files specifying the packages and versions of that project. This isolation prevents conflicts between different projects. There is also a global environment that maintains the packages and versions available throughout the system. Julia also provides facilities for the development of packages. Developers can create their own packages, publish them and modify the package code. Julia allows working on packages in a mode called 'development mode'. This mode is useful for debugging and improving packages.

1.3.3 Development Environments

Julia is said to provide different development environments and scientific computability features. Julia has a large package ecosystem that allows developers to create and publish their own packages. It also provides the possibility to modify the package code and test it interactively by working in development mode.

1.3.4 Scientific and Computational Environments

Julia can operate in a collaborative environment using distributed computing environments and tools such as JupyterHub. It can be run on cloud services, e.g. AWS, Google Cloud, and can also be containerised using Docker, enabling the creation of scalable and repeatable computing environments. These different environments provided by Julia support a variety of workflows and are versatile for both research and production uses.

In summary, Julia provides a variety of environments for different use cases, from interactive development and testing to managing project dependencies and deploying applications. These environments support a range of workflows and make Julia versatile for both research and production use.

Julia offers many powerful packages for data science and machine learning. Julia has an impressive set of packages for data processing, visualisation, deep learning, machine learning and optimisation.

- ***DataFrames.jl***, is a data processing library that makes it easy to work with tabular data.
  DataFrames.jl is a popular package for data manipulation and analysis in Julia. It provides similar functionality to that found in pandas for Python or data.table for R, allowing efficient operations on data tables.
  You can use Julia's package manager to install DataFrames.jl. You can open Julia REPL and write;

using Pkg
Pkg.add('DataFrames')
command lines. For more information, it is recommended to visit https://dataframes.juliadata.org/stable/.

- ***Plots.jl***, is a flexible and feature-rich library for data visualisation.
  Plots.jl is a powerful and flexible plotting package that provides a unified interface for various backends in Julia programming language, allowing users to create a wide variety of plots. It is highly customisable and can work in harmony with other Julia packages.
  To install the Plots.jl file, the Julia REPL must be opened and using the package manager;
  using Pkg
  Pkg.add('Plots')
  command lines must be typed. For more information, it is recommended to visit https://docs.juliaplots.org/stable/.
- ***Flux.jl***, stands out as a deep learning library that supports deep learning techniques.
  Flux.jl is a powerful and flexible machine learning library for Julia, designed for building and training neural networks. It offers a wide range of tools for deep learning, from simple models to complex neural network architectures.
  To install the Flux.jl file, the Julia REPL must be opened and using the package manager;
  using Pkg
  Pkg.add("Flux")
  command lines must be typed. For more information, it is recommended to visit https://fluxml.ai/Flux.jl/stable/.
- ***MLJ.jl***, is a machine learning framework that allows working with various machine learning models with a user-friendly interface.
  MLJ.jl is a comprehensive machine learning library for Julia that provides a unified interface for a wide range of machine learning tasks, including model training, evaluation and hyperparameter settings. It integrates well with other Julia packages and provides tools for data preprocessing, model selection and performance evaluation.
  To install the MLJ.jl file, the Julia REPL must be opened and using the package manager;
  using Pkg
  Pkg.add("MLJ")
  command lines must be typed. For more information, it is recommended to visit https://juliaai.github.io/MLJ.jl/stable/.
- ***JuliaDB.jl***, is a package used to work with datasets too large to fit in memory.
  JuliaDB.jl is a Julia package designed for high performance data processing and storage, especially suitable for handling large datasets. It offers features similar to those found in other data management systems such as SQL databases but is optimised for the Julia programming language.

To install the JuliaDB.jl file, the Julia REPL must be opened and using the package manager;

using Pkg

Pkg.add("JuliaDB")

command lines must be typed. For more information, it is recommended to visit https://juliadb.juliadata.org/latest/.

- ***Optim.jl,*** is a powerful optimisation library that solves a wide range of optimisation problems.
  Optim.jl is a Julia package designed for numerical optimisation. It offers various algorithms for solving optimisation problems, including unconstrained and constrained optimisation for both smooth and non-uniform problems. The package is suitable for tasks such as finding local minima, maximising functions and solving non-linear least squares problems.

  To install the Optim.jl file, the Julia REPL must be opened and using the package manager;

  using Pkg

  Pkg.add("Optim")

  command lines must be typed. For more information, it is recommended to visit https://docs.sciml.ai/Optimization/stable/optimization_packages/optim/.
- ***ScikitLearn.jl***, For machine learning.
  ScikitLearn.jl is a Julia package that provides an interface to Python's scikit-learn, allowing you to use scikit-learn machine learning algorithms and tools directly in Julia. This package leverages Julia's PyCall.jl package to interface with the Python ecosystem, allowing scikit-learn functionality to be seamlessly integrated into Julia workflows.

  To use ScikitLearn.jl, you need to have Python and scikit-learn installed on your system.

  To install the ScikitLearn.jl file, the Julia REPL must be opened and using the package manager;

  using Pkg

  Pkg.add("ScikitLearn")

  command lines must be typed. For more information, it is recommended to visit https://cstjean.github.io/ScikitLearn.jl/dev/man/models/.
- ***JuMP.jl,*** For optimisation.
  JuMP.jl is a powerful and flexible Julia program package for mathematical optimisation. It provides a high-level syntax for defining and solving optimisation problems and is widely used for linear programming (LP), mixed integer programming (MIP), quadratic programming (QP) and more complex optimisation tasks.

  To install the Flux.jl file, the Julia REPL must be opened and using the package manager;

  using Pkg

  Pkg.add("JuMP")

  command lines must be typed. For more information, it is recommended to visit https://jump.dev/JuMP.jl/stable/.

- ***DifferentialEquations.jl***, For solving differential equations.
  DifferentialEquations.jl is a Julia package for solving various types of differential equations. It offers a flexible and high-performance interface for solving ordinary differential equations (ODEs), partial differential equations (PDEs), stochastic differential equations (SDEs) and delay differential equations (DDEs). The package supports a large number of numerical methods and solution tools, making it a powerful tool for both academic research and practical applications.
  To install the DifferentialEquations.jl file, the Julia REPL must be opened and using the package manager;
  using Pkg
  Pkg.add("DifferentialEquations")
  command lines must be typed. For more information, it is recommended to visit https://docs.sciml.ai/DiffEqDocs/stable/.
- ***Optimisation.jl***, For optimisation problems.
  Optimisation.jl is a Julia package designed to provide a comprehensive suite of tools for optimisation problems. It focuses on providing a unified interface to various optimisation solvers and algorithms for different types of optimisation, including linear programming, quadratic programming and non-linear optimisation.
  To install the Optimisation.jl file, the Julia REPL must be opened and using the package manager;
  using Pkg
  Pkg.add("Optimisation")
  command lines must be typed. For more information, it is recommended to visit https://docs.sciml.ai/Optimisation/stable/getting_started/.
- ***LinearAlgebra.jl***, is used for standard linear algebra operations.
  LinearAlgebra.jl is a comprehensive Julia package that provides a wide range of functions for linear algebra operations. It includes tools for matrix and vector calculations, solving systems of linear equations, eigenvalue problems, singular value decomposition and many more. This package is a fundamental part of Julia's scientific computing ecosystem and is widely used in many applications.
  Since LinearAlgebra.jl is part of Julia's standard library, it comes preinstalled with Julia and does not require a separate installation. For more information, it is recommended to visit https://docs.julialang.org/en/v1/stdlib/LinearAlgebra/.
- ***StaticArrays.jl***, For small fixed-size arrays with high performance.
  StaticArrays.jl is a Julia package designed for the performant handling of small, fixed-size arrays and matrices. Unlike dynamically sized arrays, StaticArrays.jl arrays have a fixed size known at compile time, which makes it easier to perform optimisations and can lead to performance improvements, especially in numerical computing and scientific applications.
  To install the StaticArrays.jl file, the Julia REPL must be opened and using the package manager;
  using Pkg
  Pkg.add("StaticArrays")

command lines must be typed. For more information, it is recommended to visit https://juliaarrays.github.io/StaticArrays.jl/stable/.

- ***Makie.jl***, For high-performance and interactive visualisations.
  Makie.jl is a powerful and flexible drawing library for Julia, designed to create high-performance, interactive visualisations. It offers a wide range of capabilities for 2D and 3D drawing. It is well known for its flexibility and high-quality rendering. Makie.jl is well integrated with Julia's numerical and scientific computing ecosystem.
  To install the Makie.jl file, the Julia REPL must be opened and using the package manager;
  using Pkg
  Pkg.add("Makie")
  command lines must be typed. For more information, it is recommended to visit https://docs.makie.org/v0.21/.
- ***http.jl***, For HTTP client and server functionality.
  HTTP.jl is a Julia package for handling HTTP requests and responses. It provides an easy-to-use interface for creating HTTP clients and servers, making it suitable for applications such as web scraping, RESTful API interactions, and web services creation and development.
  To install the http.jl file, the Julia REPL must be opened and using the package manager;
  using Pkg
  Pkg.add("HTTP")
  command lines must be typed. For more information, it is recommended to visit https://juliaweb.github.io/HTTP.jl/v0.6/.
- ***Genie.jl***, A web framework for building web applications.
  Genie.jl is a fully comprehensive web framework for Julia, designed to make web development simple and efficient. It offers a variety of tools for building web applications, APIs and microservices based on performance and ease of use.
  Key Features are; MVC Architecture: Supporting Model-View-Controller (MVC) architecture for organising code, Built-in ORM: An Object Relational Mapping (ORM) system for database interactions, Routing and Middleware: Routing and middleware support for handling web requests, Drafting Infrastructure: Providing routing and middleware support for processing web requests.
  To install the Genie.jl file, the Julia REPL must be opened and using the package manager;
  using Pkg
  Pkg.add("Genie")
  command lines must be typed. For more information, it is recommended to visit https://genieframework.com.

1.4 Installation and Julia Packages

Regarding the installation of Julia, it is stated that you should download the latest version from Julia's website (Download Julia (julialang.org)) [6] and install it according to your operating system. The steps on how to install the downloaded files are also explained in detail. It is also recommended to add Julia to PATH to enable you to run Julia from the command line.

To verify that the installation was done correctly, you should run Julia from Terminal or command prompt and verify the startup and version information.

Regarding Julia packages, it introduces Pkg, Julia's built-in package manager. This package manager allows you to add, remove and update packages, as well as manage dependencies.

1.4.1 Download and Install Julia [7, 8]

- Download Julia:
  - Visit the official Julia website to download the latest stable version of Julia.
  - Choose the appropriate version for your operating system.
  - Click on the link to download the installer. For Windows, this will usually be an.exe file; for macOS, a.dmg file; and for Linux, a.tar.gz file.
- Install Julia:

Windows:

- Run the installer.exe file and follow the installation prompts.
- Ensure that you check the option to add Julia to your system PATH.

macOS:

- Open the downloaded.dmg file and drag the Julia application to your Applications folder.
- Add Julia to your PATH (it is optional).
- Open terminal and run *echo 'export PATH = "/Applications/Julia-<version>/Contents/Resources/julia/bin:$PATH"' >>~/.zshrc* ommand. (Replace <version> with your Julia version number. If you use Bash, modify ~/.bash_profile instead.)

Linux:

- Extract the downloaded.tar.gz file to a directory of your choice and run the Julia executable from that directory.
- Run *tar-xvzf julia-<version>-linux-x86_64.tar.gz* command. (Replace <version> with the version you downloaded.)

1.4.2 Verify Installation

- Open the Julia REPL (Read-Eval-Print Loop) by running julia in your terminal or command prompt.
- Verify that Julia starts up and displays its version information.

1.4.3 Installing Julia Packages [8]

1. *Package Manager Overview*, Julia uses its built-in package manager called Pkg for managing packages. The package manager allows you to add, remove, and update packages, and manage dependencies.
2. *Basic Package Management Commands*
    - Opening Julia in REPL,
    - Enter the Pkg mode (a command-line interface for package management).
    - Install a package: For example, to install the DataFrames package, type: *add DataFrames* command.
    - Exit Pkg REPL mode, Press Backspace or Ctrl+C to return to the Julia prompt.
3. Updating PackagesOpen Julia REPL.
    - Open Julia REPL.
    - Enter Pkg REPL mode.
    - Update all packages by using *update* command.
    - To update a specific package, use *update PackageName* command.
4. *Removing Packages*
    - Open Julia REPL.
    - Enter Pkg REPL mode.
    - Remove a package: For example, to remove the DataFrames package, type *rm DataFrames* command.

References

1. R. Gökhan Türeci, Hamdi Dağıstanlı, İlkay Türk Çakır. Fizik ve Mühendislikte Python. ISBN: 9786258440328, Gazi Kitabevi, 2021
2. Sandeep, Nagar, Beginning Julia Programing For Engineers and Sceintist. APRESS, ISBN-13 (pbk): 978-1-4842-3170-8 ISBN-13 (electronic): 978-1-4842-3171–5, https://doi.org/10.1007/978-1-4842-3171-5, 2017
3. The Julia Language, Julia V.1.9.3, 2023
4. Why Use Julia? 12 Benefits of Julia Programming – Master Data Skills + AI (enterprisedna.co)

5. The Julia Programming Language: The History and Uses (leftronic.com)
6. Download Julia (julialang.org)
7. Getting Started · The Julia Language
8. Installation · The Julia Language

Chapter 2
Variables and Operators

Julia provides a wide range of basic numeric types on which a full complement of standard mathematical functions as well as arithmetic and bitwise operators are defined. These map directly to numeric types and operations natively supported on modern computers, thus allowing Julia to take full advantage of computational resources. In addition, Julia provides software support for Arbitrary Precision Arithmetic, which can perform operations on numerical values that cannot be represented efficiently.

A variable is a named storage location in memory that holds a value that can be changed during the execution of the programme. In the Julia programming language, the term '*variable*' refers to a symbolic name associated with a value stored in memory. The value associated with a variable can change or remain constant during the execution of the programme [1].

There are different data types used effectively in the Julia programming language. These data types and their properties can be listed as follows [2]:

int (integer) [integer]	A data type defined by integers.
float [Decimal]	It is the data type that defines decimal numbers.
str (string) [String]	A data type consisting of combinations of letters or numbers.
bool [0 and 1]	It is a data type with True and False values.
Array	An array is a data structure that can hold a collection of elements, all of which can be of the same type (though Julia also allows heterogeneous arrays). Arrays are versatile and can be one-dimensional (vectors), two-dimensional (matrices), or multi-dimensional.
tuple [Bunch]	This data type is very similar to the list data type. But the difference from the list data type is that it cannot be modified.

R. G. Türeci et al., *Julia Programming for Physics Applications*,
https://doi.org/10.1007/978-3-031-84716-5_2

2.1 Number Types

2.1.1 Integer Numbers

Integers are whole numbers without fractional parts. Julia supports both signed and unsigned integer types, which vary in the range of values they can represent and whether they can represent negative numbers.

Julia has many types of integers, which vary in size and whether they are signed or unsigned.

For more information about integer numbers in Julia, you can visit Julia documentation-types web page https://docs.julialang.org/en/v1/manual/types/ [3].

Signed Integers:

Int8	:	8-bit signed integer, ranging from −128 to 127
Int16	:	16-bit signed integer, ranging from −32,768 to 32,767
Int32	:	32-bit signed integer, ranging from −2,147,483,648 to 2,147,483,647
Int64	:	64-bit signed integer, ranging from −9,223,372,036,854,775,808 to 9,223,372,036,854,775,807
Int128	:	128-bit signed integer, with an extremely large range

Unsigned Integers:

UInt8	:	8-bit unsigned integer, ranging from 0 to 255
UInt16	:	16-bit unsigned integer, ranging from 0 to 65,535
UInt32	:	32-bit unsigned integer, ranging from 0 to 4,294,967,295
UInt64	:	64-bit unsigned integer, ranging from 0 to 18,446,744,073,709,551,615
UInt128	:	128-bit unsigned integer, with an extremely large range

2.1.2 Floating-Point Types

Floating-point numbers are real numbers with fractional parts. Julia uses the IEEE 754 standard for floating-point arithmetic.

Floating-point numbers are real numbers with fractional parts. Julia uses the IEEE 754 standard for floating-point arithmetic. There are four types of floating-point numbers;

Single-Precision Float:

Float32	:	32-bit single-precision floating-point, with approximately 7 decimal digits of precision
Double-Precision Float: Float64	:	64-bit double-precision floating-point, with approximately 15 decimal digits of precision. This is the default floating-point type in Julia
Extended-Precision Float: Float16	:	16-bit half-precision floating-point, with less precision and range than Float32. Typically used in machine learning contexts
Arbitraru-Precision Float: BigFloat	:	Arbitrary-precision floating-point numbers. Typically used in computations requiring very high precision

For more information about floating-points types in Julia, you can visit Julia documentation-types web page https://docs.julialang.org/en/v1/manual/types/ [4].

2.1.3 Complex Types

Complex numbers have a real and imaginary part, both of which are floating-point numbers. Computations involving complex numbers are used in many areas of science and mathematics. Complex numbers provide a broader description of physical phenomena than real numbers. The Julia language can easily handle complex numbers to successfully perform mathematical calculations. Complex numbers are defined in a simple way and can be used with various functions. Furthermore, their use in combination with other data types is flexible and seamless. These features make Julia one of the best choices for complex number operations and analyses

Complex{T}	:	Denotes x complex number of type T with real and imaginary parts. For example, Complex{Float32} denotes x complex number with both real and imaginary parts Float32

The Programme 2.1, shows the outputs of the real and imaginary parts of a given complex number. The real part of x complex number is -2 and the imaginary part is 5i. In Julia, complex numbers are stored as two numbers, re (the real part) and im (the imaginary part). Both of these are a kind of real number. They can be obtained using the *real()* and *imag()* functions.

Programme 2.2, shows the inputs of the real and imaginary parts and the complex number obtained by using the input values. The complex number can also be formed by defining the real and imaginary parts as separate variables.

Programmes 2.3 and 2.4 shows the sum, subtraction, multiplication, division, and power of two matrices by using complex numbers.

```
[9]:
x=-2+5im
[9]:
-2 + 5im
[11]:
real(x)
[11]:
-2
[13]:
imag(x)
[13]:
5
```

Programme 2.1 Definition of real and integer part of a complex number

```
x=4
[15]:
4
[17]:
y=-5
[17]:
-5
[19]:
z=complex(x,y)
[19]:
4 - 5im
```

Programme 2.2 Definition of real and integer part of a complex number

```
[21]: x=2.3
[21]: 2.3
[23]: y=-4.5
[23]: -4.5
[25]: z=complex(x,y)
[25]: 2.3 - 4.5im
[27]: z1=-z
[27]: -2.3 + 4.5im
[29]: ztot=z+z1
[29]: 0.0 + 0.0im
[31]: zmin=z-z1
[31]: 4.6 - 9.0im
[33]: zmul=z*z1
[33]: 14.96 + 20.7im
[35]: zdiv=z/z1
[35]: -0.9999999999999999 - 0.0im
[39]: zexp=z^z1
[39]: -3.1181395185276264 - 1.2885253529711844im
```

Programme 2.3 Addition, subtraction, multiplication, division and power of two matrices

```
[41]: zpow=z^2
[41]: -14.96 - 20.7im
[43]: zsca=-3*z
[43]: -6.8999999999999995 + 13.5im
[45]: ztut=z*(zpow/zmul-zsca)
[45]: -47.18 - 57.599999999999994im
```

Programme 2.4 Power and multiplication with a real number of a matrix

The complex number can also be generated using different types of data.

Programme 2.5 shows the types of imaginary and reel part of a matrix by using different data types.

For more information about complex types in Julia, you can visit Julia documentation-types web page https://docs.julialang.org/en/v1/manual/types/ [5].

```
[47]: x=Int8(75)

[47]: 75

[55]: y=Float32(2.3456789)

[55]: 2.3456788f0

[57]: c=complex(x,y)

[57]: 75.0f0 + 2.3456788f0im

[61]: typeof(real(c))

[61]: Float32

[63]: typeof(imag(c))

[63]: Float32
```

Programme 2.5 Mix data example

2.1.4 Rational Numbers

Rational numbers are exact fractions. They are represented as the ratio of two integers.

Rational{T} : Represents a rational number where T is the type of the numerator and denominator. For example, Rational{Int} represents a rational number whose numerator and denominator are integers

For more information about rational numbers in Julia, you can visit Julia documentation-types web page https://docs.julialang.org/en/v1/manual/types/ [6].

2.1.5 Special Numeric Types

Special numeric types can be used in two different ways: Arbitrary-precision integer and arbitrary-precision rational.

For arbitrary-precision integer, we use BigInt.

BigInt : Arbitrary-precision integers are suitable for calculations with very large integers that exceed the range of the default integer type

For arbitrary-precision rational, we use BigRat.

BigRat : It is used for rational numbers of arbitrary precision, useful for exact arithmetic involving large numerators and denominators

For more information about special numeric types in Julia, you can visit Julia documentation-types web page https://docs.julialang.org/en/v1/manual/types/ [7].

2.1.6 Numeric Literal Types

Julia allows numeric literals in different formats. In Julia, numeric literals can be of different types depending on their form and the context in which they are used. As an example of these usage contexts, integer, floating-point, complex and special numeric literals are explained below with examples according to the context in which they are used.

1. Integer Literals

Binary	:	0b1010 is interpreted as an Int (with base 2)
Octal	:	0o52 is interpreted as an Int (with base 8)
Hexadecimal	:	$0 \times 2A$ is interpreted as an Int (with base 16)

2. Floating Point Literals

Decimal	:	3.14 is interpreted as Float64 by default
Scientific Notation	:	1.23e4 is also interpreted as a Float64

3. Complex Literals
   Complex: $1 + 2im$ is interpreted as Complex{Int} or Complex{Float64} depending on the types of the real and imaginary parts. If one of the parts is Float64, the complex number will be of type Complex{Float64}.
4. Special Numeric Literals

   Large Integers: numbers that exceed the default Int range are treated as BigInt.

For more information about numerical literal types in Julia, you can visit Julia documentation-types web page https://docs.julialang.org/en/v1/manual/types/ [8].

2.2 Data Types

Julia does not have a specific data type called 'list', as you might find in some other programming languages such as Python. Instead, Julia uses various data structures to handle collections of items. The closest equivalents to lists in Julia are:

In Julia, there isn't a specific data type named 'list' as you might find in some other programming languages like Python. Instead, Julia uses a variety of data structures to handle collections of elements. The closest equivalents to lists in Julia are arrays, and tuples.

For more information about data types in Julia, you can visit Julia documentation-types web page https://docs.julialang.org/en/v1/manual/types/ [9].

2.2.1 Arrays

Arrays in julia are the most common method of representing a list of items. They are ordered, mutable, and can store items of the same type. For example:

Programme 2.6 defines 1D array whose elements are integers, Programmes 2.7 and 2.8 define 1D and 2D arrays whose elements are strings respectively.

You can create arrays with different types, access elements using indices, and use various In the Julia programming language, different types of arrays can be created, elements can be accessed by using indices and various functions can be used to process this data.

For more information about arrays in Julia, you can visit Julia documentation-arrays web page https://docs.julialang.org/en/v1/base/arrays/ [10].

```
[5]: arr=[1,2,3,4,5,6]#defining 1D array of integers

[5]: 6-element Vector{Int64}:
      1
      2
      3
      4
      5
      6
```

Programme 2.6 Defining 1D array of integers

```
[9]: arr = ["one", "two", "three","four"]#defining 1D array of strings

[9]: 4-element Vector{String}:
      "one"
      "two"
      "three"
      "four"
```

Programme 2.7 Defining 1D array of strings

```
[17]: m = [1 2 3 ; 5 6 7; 8 9 10 ; 11 12 13]#defining 2D array

[17]: 4×3 Matrix{Int64}:
        1   2   3
        5   6   7
        8   9  10
       11  12  13
```

Programme 2.8 Defining 2D array of strings

```
[19]:  t=(1,2,3,4,5,6)#defining a tuple of integers

[19]:  (1, 2, 3, 4, 5, 6)
```

Programme 2.9 Defining a tuple of integers

```
[21]:  tm=(1,"julia",4.56,2^3)#defining a tuple with mixed types

[21]:  (1, "julia", 4.56, 8)
```

Programme 2.10 Defining a tuple with mixed types

2.2.2 Tuples

Tuples are collections of items that cannot be modified. Unlike arrays, once a tuple is created, its size and contents cannot be changed. Tuples are used when you want to operate on a fixed size collection. For example.

Programme 2.9 defines a tuple whose elements are integers and Programme 2.10 defines a tuple whose elements are mixed types.

For more information about tuples in Julia, you can visit Julia documentation-types web page https://docs.julialang.org/en/v1/manual/types/ [11].

2.2.3 Lists in Julia Via Linked Lists

While Julia does not have a built-in linked list type, you can implement linked lists or use packages that provide them.

2.2.4 Custom Data Structures

For specialised use cases, you might want to define your own data structures. Julia allows you to create custom types using ‘struct’. For example:

Programme 2.11 defines a custom structure named *MyL1* and Programme 2.12 defines elements of the list.

For more information about lists in Julia, you can visit Julia documentation-types web page https://docs.julialang.org/en/v1/manual/types/ [12].

```
[39]: struct MyL1
          elements::Vector{Int}
      end

[ ]:
```

Programme 2.11 Defining a custom structure

```
[39]: struct MyL1
          elements::Vector{Int}
      end

[43]: myfl = MyL1([1, 2, 3, 4, 5])

[43]: MyL1([1, 2, 3, 4, 5])
```

Programme 2.12 Creating a list

```
[45]: set1 = Set([1, 2, 3, 4, 5, 6])

[45]: Set{Int64} with 6 elements:
        5
        4
        6
        2
        3
        1
```

Programme 2.13 Creating a set

2.2.5 Sets

Although not ordered like lists, sets are collections of unique elements. For example:

Programme 2.13 shows the creation of a set.

For more information about sets in Julia, you can visit Julia documentations-collections and data structures web page https://docs.julialang.org/en/v1/base/collections/ [13].

2.2.6 LinkedList Package

For more complex linked list structures, the 'DataStructures' package can be used, which provides a 'LinkedList' type.

Programme 2.14 defines using of DataStructures package.

```
[49]: using DataStructures

[53]: MyL1 = LinkedList(5, 10, 15, 20, 25, 30, 35, 40)
```

Programme 2.14 Creating a linked list using the DataStructures packages

```
[55]: x_true=true

[55]: true

[57]: x_false=false

[57]: false

[59]: typeof(x_true)

[59]: Bool
```

Programme 2.15 Definition of bool data type

For more information about lists in Julia, you can visit Lists.jl, web page https://juliapackages.com/p/lists [14].

2.3 Bool Data Type

In Julia, the 'Bool' data type represents boolean values that can be either 'true' or 'false'. Here are some examples of how the 'Bool' type works and how you can use it in Julia.

Programme 2.15 shows the definition of a bool data type.

2.4 Boolean Operations

You can use standard logical operations with boolean values:

Logical AND	:	&& or &
Logical OR	:	|| or |
Logical NOT	:	!
XOR	:	⊻

Programme 2.16 shows using of logical operators and their results.

Booleans are often used in conditional expressions such as 'if', 'while' and 'for' loops. See Chap. 3 for the use of logical operators in if, for and while loops.

```
[61]: o1 = true && false # using && operator
[61]: false
[63]: o2 = true & true # using & operator
[63]: true
[65]: o3 = true || false # using || operator
[65]: true
[67]: o4 = false | false # using | operator
[67]: false
[69]: o5 = !true # using ! operator
[69]: false
[71]: o6 = !false # using ! operator
[71]: true
[73]: o7 = true ⊻ false # using ⊻ operator
[73]: true
[75]: o8 = true ⊻ true # using ⊻  operator
[75]: false
```

Programme 2.16 Using logical operators

You can convert between Boolean values and other types:

Programme 2.17 indicates the conversion between boolean and integer values.

By using command lines shown in Programme 2.17, you can convert integer values to boolean values. '0' will convert to 'false' and any non-zero integer will convert to 'true'.

Programme 2.18 indicates the conversion between boolean and string values.

```
[1]: bool_zero = Bool(0)
[1]: false
[3]: bool_non_zero = Bool(1)
[3]: true
```

Programme 2.17 Conversion between Boolean values and integer values

```
[*]: bool_str = parse(Bool, 'true')
```

Programme 2.18 Conversion between Boolean values and string values

By using command lines shown in Programme 2.18, you can convert string values to boolean values. You can use 'parse' with 'Bool', but it is not recommended.

For more information about boolean operators in Julia, you can visit Julia Manual-Mathematical Operations and Elementary Functions web page https://docs.julialang.org/en/v1/manual/mathematical-operations/ [15].

2.5 Matrices and Matrix Algebra

Julia is well suited for linear algebra operations and matrix operations, being a powerful programming language designed for high-performance numerical and scientific computing.

In order to perform linear algebra operations in the Julia programming language, the *LinearAlgebra* module must be imported using the *using LinearAlgebra* command line.

In Julia programming language, matrices are created in square brackets and operations are performed on these matrices.

Programme 2.19 indicates how to define matrices in Julia. The lines of this command indicate that matrix A is a 2 × 3 matrix and matrix B is a 2 × 3 matrix. By typing these commands, matrices A and B will be defined.

After defining the matrices, operations related to matrices A and B can be done very easily.

Programme 2.20 defines the sum and the subtraction of two matrices. By typing these command lines, matrix C, which is the sum of matrices A and B, and matrix D, which is the difference of matrices A and B, will be calculated.

In addition to addition and subtraction operations in matrices, multiplication of matrices by a scalar and scalar multiplication of two matrices can also be performed.

If it is desired to multiply a matrix by a scalar;

it will be enough to type the command line as shown in Programme 2.21. Programme 2.21 defines the multiplication of matrix by a scalar.

If two matrices are to be multiplied with each other, matrix multiplication is performed with the * operator. Multiplication of two matrices can be performed by

```
[11]: A = [1 2 3 4; 3 4 5 6]

[11]: 2×4 Matrix{Int64}:
       1  2  3  4
       3  4  5  6

[13]: B = [5 6 7 8; 7 8 9 10]

[13]: 2×4 Matrix{Int64}:
       5  6  7   8
       7  8  9  10
```

Programme 2.19 Defining matrices in Julia

```
[17]: C=A+B

[17]: 2×4 Matrix{Int64}:
        6   8  10  12
       10  12  14  16

[19]: D=A-B

[19]: 2×4 Matrix{Int64}:
       -4  -4  -4  -4
       -4  -4  -4  -4
```

Programme 2.20 Matrices operations in Julia

```
[23]: Mult=3*B

[23]: 2×4 Matrix{Int64}:
       15  18  21  24
       21  24  27  30
```

Programme 2.21 Multiplication a matrix by a scalar

typing the command line as shown in Programme 2.22. Programme 2.22 defines the multiplication of two matrices.

Among the operations to be used in calculations on matrices, in general, the transpose, determinant, inverse of the matrix and finding eigenvalues and eigenvectors are included.

```
[31]: m_1 = [2 3; 3 4]

[31]: 2×2 Matrix{Int64}:
       2  3
       3  4

[33]: m_2=[4 5; 6 7]

[33]: 2×2 Matrix{Int64}:
       4  5
       6  7

[37]: mult=m_1*m_2

[37]: 2×2 Matrix{Int64}:
       26  31
       36  43
```

Programme 2.22 Multiplication of two matrices

For taking the tranpose of a matrix; command lines can be used as shown in Programme 2.23.

To compute the determinant of a square matrix, command lines in Programme 2.24 can be used.

For computing the inverse of a matrix, command lines in Programme 2.25 can be used.

For computing eigenvalues, eigenvectors and symmetry of a matrix, command lines in Programme 2.26 can be used.

```
[39]: m_3 = transpose(m_1)

[39]: 2×2 transpose(::Matrix{Int64}) with eltype Int64:
       2  3
       3  4

[41]: m_3=m_1'

[41]: 2×2 adjoint(::Matrix{Int64}) with eltype Int64:
       2  3
       3  4
```

Programme 2.23 Transpose of a matrix

```
[5]: m_1 = [2 3; 3 4]

[5]: 2×2 Matrix{Int64}:
      2  3
      3  4

[9]: using LinearAlgebra

[11]: det(m_1)

[11]: -1.0000000000000004
```

Programme 2.24 Determinant of a square matrix

```
[3]: m_1 = [2 3; 3 4]

[3]: 2×2 Matrix{Int64}:
      2  3
      3  4

[49]: inv(m_1)

[49]: 2×2 Matrix{Float64}:
       -4.0   3.0
        3.0  -2.0
```

Programme 2.25 Inverse of a square matrix

```
[13]: m_1 = [2 3; 3 4]

[13]: 2×2 Matrix{Int64}:
       2  3
       3  4

[15]: using LinearAlgebra

[19]: eigvals(m_1)

[19]: 2-element Vector{Float64}:
      -0.1622776601683793
       6.16227766016838

[23]: eigvecs(m_1)

[23]: 2×2 Matrix{Float64}:
      -0.811242  0.58471
       0.58471   0.811242

[25]: sm1 = Symmetric(m_1)

[25]: 2×2 Symmetric{Int64, Matrix{Int64}}:
       2  3
       3  4
```

Programme 2.26 Eigenvalues, eigenvectors and symmetry of a square matrix

Julia's 'LinearAlgebra' module provides a rich set of tools for working with matrices and performing linear algebra operations efficiently. For further information on matrix operations and for advanced features, it is recommended to consult Julia-Standart Library-Linear Algebra web page https://docs.julialang.org/en/v1/stdlib/LinearAlgebra/ [16].

References

1. Sandeep, Nagar, Beginning Julia Programing For Engineers and Scientist. APRESS, ISBN-13 (pbk): 978-1-4842-3170-8 ISBN-13 (electronic): 978-1-4842-3171-5, https://doi.org/10.1007/978-1-4842-3171-5, 2017
2. The Julia Language, The Julia Project, Julia V.1.9.3, 2023
3. Julia documentation-Types-Integer numbers. https://docs.julialang.org/en/v1/manual/types/
4. Julia documentation-Types-Floating-points types. https://docs.julialang.org/en/v1/manual/types/
5. Julia documentation-Types-Complex types. https://docs.julialang.org/en/v1/manual/types/
6. Julia documentation-Types-Rational numbers. https://docs.julialang.org/en/v1/manual/types/
7. Julia documentation-Types-Special numeric types. https://docs.julialang.org/en/v1/manual/types/
8. Julia documentation-Types-Literal types. https://docs.julialang.org/en/v1/manual/types/
9. Julia documentation-Types-Data types, https://docs.julialang.org/en/v1/manual/types/
10. Julia documentation-Arrays, https://docs.julialang.org/en/v1/base/arrays/
11. Julia documentation-Types , https://docs.julialang.org/en/v1/manual/types/
12. Julia documentation-Types, https://juliapackages.com/p/lists

13. Julia documentations-collections and data structures, https://docs.julialang.org/en/v1/base/collections/
14. Lists.jl, https://juliapackages.com/p/lists
15. Julia manual-Mathematical Operations and Elementary Functions. https://docs.julialang.org/en/v1/manual/mathematical-operations/
16. Julia-Standart Library-Linear Algebra. https://docs.julialang.org/en/v1/stdlib/LinearAlgebra/

Chapter 3
Loops and Conditional Situations

3.1 Introduction

In this section, one of the most important features of programming, loops and conditionals, will be discussed. Repeated operations from an initial value to a final value are performed with loops. For example, if you want to calculate the sum of even numbers from 1 up to and including 100, looping is a practical way. Conditional situations allow a loop to terminate when the condition is met, or to continue the loop until the condition is met. For a slightly more challenging example, consider an example that requires both a conditional and a loop: In an application that wants to calculate the sum of prime numbers from 1 to 1000, a loop is required. This loop first checks whether the number is prime or not and if the number is prime it is stored in a list. When the number 1000 is reached, the sum of the list elements can be calculated, and the question can be given easily.

The determination of Fourier series in the analysis of a sound or an electrical signal, or the expansion of a function into a Legendre series can also be performed with a conditional loop. Root-finding algorithms such as Newton-Raphson or Muller's method, which can be used to determine the root of a function, also require conditional loops. Therefore, loops are an extremely important argument in programming.

3.2 in operator

In operator allows to query a specific number in an array of numbers, which is a number in a list data type, or similarly a number in a tuple data type, or a string data type, or dictionary data type. If the queried quantity is present in the object, the result returns as ***true***; if not, the result returns as ***false***.

R. G. Türeci et al., *Julia Programming for Physics Applications*,
https://doi.org/10.1007/978-3-031-84716-5_3

In Programme 3.1, the in operator is applied to the list ***L***. The output of the first cell is *5-element Vector(Int64)* indicating that the list is an integer array of 5 elements. The second cell asks whether the number 9 is in the list and the third cell asks whether the number 11 is in the list. The second cell returns as ***true***, and the third cell returns as ***false***. The in operator can also be applied to the set data type, see Programme 3.2. Programme 3.3 represents the application of ***in*** operator to tuple data type.

```
[1]:  1 L = [1, 3, 5, 7, 9]

[1]:  5-element Vector{Int64}:
       1
       3
       5
       7
       9

[2]:  1 9 in L

[2]:  true

[3]:  1 11 in L

[3]:  false
```

Programme 3.1 The *in* operator for *list* data type

```
[4]:  1 S = ([1, 3, 5, 7, 9])

[4]:  5-element Vector{Int64}:
       1
       3
       5
       7
       9

[5]:  1 9 in S

[5]:  true
```

Programme 3.2 The *in* operator for *set* data type

```
[6]:  1 T = (2, 4, 6, 8, 10)

[6]:  (2, 4, 6, 8, 10)

[7]:  1 10 in T

[7]:  true

[8]:  1 1 in T

[8]:  false
```

Programme 3.3 The *in* operator for *tuple* data type

```
[10]: 1 s = "Julia"

[10]: "Julia"

[11]: 1 occursin("J", s)

[11]: true

[12]: 1 occursin("Heisenberg", "Werner Heisenberg")

[12]: true
```

Programme 3.4 The *occursin()* for string data type

3.2.1 occursin()

For string data type, the ***occursin()*** is used, not the in operator. This command takes two parameters. The first one is the searched substring, and the second parameter is the string itself. It is a command that can be used for data type in text structure, Programme 3.4.

3.3 The for loop

The ***for*** loop is used in conjunction with the in operator to perform a certain operation on the elements of a given object, on the elements of the corresponding data type. The for loop must end with end and the operation to be performed by the for loop must be written with a TAB space. Just like in Python, indentation is very important in Julia programming language. Another important condition is to use ^ (Shift + 3) to get the exponent of a number, $3^2 = 9$.

For example, the for loop can be used to calculate the squares of the elements of a list given from the number 1 to the number 5, Program 3.5. Here we can mention one detail. If this is done with the ***print()*** function, the calculated values are written side by side and contiguously. Using the ***println()*** function for this purpose will be useful for printing the calculated values one after the other and getting a more reasonable output. Instead of ***println(), print()*** can also be used with the "\n" parameter. With this parameter, the calculated values are printed one after the other as in the ***println()*** function.

The output of Programme 3.5 is not clear to understand. Therefore, the output should be ***println(i²)*** or ***print(i², "\n")*** functions as in Programme 3.6.

```
[1]: L = [1, 2, 3, 4, 5]

     for i in L
         print(i^2)
     end

     1491625
```

Programme 3.5 The *for* loop over a list data type

```
[2]: for i in L
         println(i^2)
     end

     1
     4
     9
     16
     25

[3]: for i in L
         print(i^2, "\n")
     end

     1
     4
     9
     16
     25
```

Programme 3.6 The *for* loop over a list

In Programme 3.7, in the first cell, the elements of the T list containing elements of different data types are printed on the screen. The T list also contains some functions that are embedded directly in the Julia language without calling a special module. For example, ***sqrt()*** calculates the square root of a number, ***log()*** is a logarithm function in base e. As in all computer programming languages, ***log()*** is a logarithm in base e. Another embedded function is the sine function. As in many programming languages, trigonometric functions calculate in radians. Therefore, the angle in degrees must be converted to radians. Another function, ***deg2rad()***, is used for this purpose. This function converts the angle value given in degrees to radians. For inverse trigonometric functions, the ***rad2deg()*** function can also be used to convert radians to degrees.

In cell 5 of the same programme, the elements of the string s are printed with a for loop. In cell 6 of the programme, the cosine values of the angles were calculated by converting these angle values to radians using the elements of the number sequence created at 30° step intervals from 0 to 180°. In the first line, the expression ***0:15:180*** is given as the ***start number***, ***step parameter*** and ***end number***. Unlike Python programming language, the last number also belongs to the number array.

The sequence of numbers created with ***0:30:180*** in cell 6 in Programme 3.7 can also be created with the ***range()*** function. The ***range()*** function is used with the ***start number***, ***end number*** and ***step parameter*** with ***step=***. This parameter is 1 by default unless the step parameter is given specifically, Programme 3.8.

In cell 12 of Programme 3.9, nested for loops are defined. The first for loop is used to count the sequence of numbers from 1 to 10, and the second for loop is used to count the symbol ~ in each round as many times as the number ***i*** in the first for loop.

In Programme 3.10, some operations were performed on the tuple data given in the list data type. In the first application, each tuple data ***i*** was printed with the counter parameter. In the next cell, the for loop is executed over the list ***L*** but over ***(i, j)***. Since the for loop is executed over ***(i, j)***, the output only gives the numbers ***i*** and ***j***.

```
[4]: T = ["Cemre", "Elif", sqrt(16), true, log(exp(2)), sin(deg2rad(20))]

     for i in T
         print(i, "\n")
     end

     Cemre
     Elif
     4.0
     true
     2.0
     0.3420201433256687

[5]: s = "Julia"

     for i in s
         println(i)
     end

     J
     u
     l
     i
     a

[6]: for i in 0:30:180
         println(cos(deg2rad(i)))
     end

     1.0
     0.8660254037844387
     0.5000000000000001
     6.123233995736766e-17
     -0.4999999999999998
     -0.8660254037844387
     -1.0
```

Programme 3.7 Some applications of *for* loop

```
[7]: for i in range(0, 180, step = 30)
         println(cos(deg2rad(i)))
     end

     1.0
     0.8660254037844387
     0.5000000000000001
     6.123233995736766e-17
     -0.4999999999999998
     -0.8660254037844387
     -1.0

[8]: for i in range(0, 1, step = 0.2)
         println(rad2deg(acos(i)))
     end

     90.0
     78.46304096718453
     66.42182152179817
     53.13010235415599
     36.86989764584401
     0.0
```

Programme 3.8 The *range()* function in *for* loop

3.4 while loop

The *while* loop is a loop structure that performs the operations in the while code block as long as the condition specified after the while word is met, and the loop terminates when the condition is not met. As in the for loop, the loop block must be

```
[9]: for i in range(0, 90, step = 15)
         println("angle = ", i, "\t", "value of sin(i)= ",sin(deg2rad(i)))
     end

     angle = 0       value of sin(i)= 0.0
     angle = 15      value of sin(i)= 0.25881904510252074
     angle = 30      value of sin(i)= 0.49999999999999994
     angle = 45      value of sin(i)= 0.7071067811865475
     angle = 60      value of sin(i)= 0.8660254037844386
     angle = 75      value of sin(i)= 0.9659258262890683
     angle = 90      value of sin(i)= 1.0

[10]: L = [i for i in range(0, 50, step = 10)]
      s = 0
      for i in L
          s += i
          println("i = ", i, "\t", "sum = ", s)
      end

      i = 0   sum = 0
      i = 10  sum = 10
      i = 20  sum = 30
      i = 30  sum = 60
      i = 40  sum = 100
      i = 50  sum = 150

[11]: sum(L)

[11]: 150

[12]: for i in 1:10
          for j in 1:i
              print("~", " ")
          end
          println()
      end

      ~
      ~ ~
      ~ ~ ~
      ~ ~ ~ ~
      ~ ~ ~ ~ ~
      ~ ~ ~ ~ ~ ~
      ~ ~ ~ ~ ~ ~ ~
      ~ ~ ~ ~ ~ ~ ~ ~
      ~ ~ ~ ~ ~ ~ ~ ~ ~
      ~ ~ ~ ~ ~ ~ ~ ~ ~ ~
```

Programme 3.9 The *for* loop with different applications

terminated with end. In Programme 3.11, a simple application where even numbers from 0 to 10 are returned is given.

Previously, in the first cell of Programme 3.9 first cell, the application in which the sine function values with 15-degree angles are calculated with for loop is shown in Programme 3.12 with while loop. ***while true*** can be used to open an infinite loop or a vicious loop in Julia. Naturally, this vicious loop is terminated with break after another condition is met. This section will be discussed after the conditional cases, i.e. *if conditional structure* is explained.

We can explain the operation of the while loop using Programme 3.13. Initially ***s = 0*** is taken. The condition for the while loop is that the value of s is equal to 1000 and the loop will run as long as ***s ≤ 1000***. In each round, the value of s is increased by 200. In the next round after s becomes 1000, the value of s becomes 1200 and the loop ends because the condition is no longer met. The value at the bottom line in the output of the programme shows this. The value of s is 1200 and the loop is terminated because the condition is no longer met.

```
[14]: L = [(1,2), (3,4), (5,6), (7,8)]
      for i in L
          println(i)
      end

      (1, 2)
      (3, 4)
      (5, 6)
      (7, 8)

[16]: for (i, j) in L
          println(i, " ", j)
      end

      1 2
      3 4
      5 6
      7 8

[18]: for (i, j) in L
          println("i :", i, "\t ", "j :", j)
      end

      i :1     j :2
      i :3     j :4
      i :5     j :6
      i :7     j :8
```

Programme 3.10 Some different *for* loop applications

```
[1]: i = 0
     while i <= 10
         println(i)
         i += 2
     end

     0
     2
     4
     6
     8
     10
```

Programme 3.11 The application of *while* loop

```
[2]: teta = 0

     while teta <= 90
         println("angle: ", teta, "\t value of sin(teta) : ", sin(deg2rad(teta)))
         teta += 15
     end

     angle: 0         value of sin(teta) : 0.0
     angle: 15        value of sin(teta) : 0.25881904510252074
     angle: 30        value of sin(teta) : 0.49999999999999994
     angle: 45        value of sin(teta) : 0.7071067811865475
     angle: 60        value of sin(teta) : 0.8660254037844386
     angle: 75        value of sin(teta) : 0.9659258262890683
     angle: 90        value of sin(teta) : 1.0
```

Programme 3.12 The value of sin function with *while* loop

```
[3]: s = 0

     while s <= 1000
         println("sum value : ", s)
         s += 200
     end
     print("the last value of s : ", s)

     sum value : 0
     sum value : 200
     sum value : 400
     sum value : 600
     sum value : 800
     sum value : 1000
     the last value of s : 1200
```

Programme 3.13 Another application of *while* loop

3.5 Conditional statements: if-else and if-elseif-else

Conditional situations are situations that we frequently encounter in daily life. A conditional statement such as "*If you get a grade of 80 or above in the exam, you can pass this course*" is used to distinguish between successful and unsuccessful students. Many applications such as checking whether a number is prime or not, calculating a Taylor series up to a certain value, interrupting and calculating the series when this certain value is met can be strongly controlled with conditional cases.

Let's start with the simplest application. First create an empty list named ***L_even*** and then create a sequence of numbers from 0 to 50 with a for loop. Since the array of numbers is created with a for block, there should be an if block in the for block so that when the remainder of the number divided by 2 is 0, this number can be appended to the list ***L_even*** with the command line ***append!(listname, element to append)***. Programme 3.14 this application is implemented (Programme 3.15).

Now Programme 3.14, let's add an else block so that the numbers that fulfil the evenness condition are stored in the *L_even* list and the numbers that do not fulfil the condition are stored in the *L_odd* list. The ***else*** code block performs exactly this function: if not.

Now let's write a programme that takes data input from the keyboard, tests whether the number entered from the keyboard is an odd number or an even number and returns the result as odd or even number.

```
[1]: L_even = []

     for i in 0:50
         if i % 2 == 0
             append!(L_even, i)
         end
     end
     print(L_even)

     Any[0, 2, 4, 6, 8, 10, 12, 14, 16, 18, 20, 22, 24, 26, 28, 30, 32, 34, 36, 38, 40, 42, 44, 46, 48, 50]
```

Programme 3.14 The even numbers from 0 to 50 with if conditional statement

```
[2]: L_even = []
     L_odd = []
     for i in 0:50
         if i % 2 == 0
             append!(L_even, i)
         else
             append!(L_odd, i)
         end
     end
     println(L_even)
     println(L_odd)

     Any[0, 2, 4, 6, 8, 10, 12, 14, 16, 18, 20, 22, 24, 26, 28, 30, 32, 34, 36, 38, 40, 42, 44, 46, 48, 50]
     Any[1, 3, 5, 7, 9, 11, 13, 15, 17, 19, 21, 23, 25, 27, 29, 31, 33, 35, 37, 39, 41, 43, 45, 47, 49]
```

Programme 3.15 If and else code block

In Julia programming language, ***readline()*** command is used to get data input from keyboard. ***parse(Int, readline(stdin))*** command line is explained:

- *stdin*: standard input,
- ***readline(stdin)***: recognises that the user has entered a line of data and recognises it as a string,
- *parse*: converts the value entered as ***stdin()*** to the specified type.

Therefore, the command line receives data input with *stdin*, this data is a string and is converted to an integer number with ***parse(Int)*** (Programme 3.16).

If the condition in the ***if*** block is not met, the condition in the ***elseif*** block is queried, if this condition is not met and there is another condition, the next elseif code block is passed to the next elseif code block, and finally the last condition is determined with the else block. Programmes 3.17 and 3.18 show the code that converts a student's grade into a letter grade. Programme 3.19 shows the code that

```
[4]: print("enter a number: ")
     n = parse(Int, readline(stdin))
     print("entered number : ", n)
     println()

     leven = []
     lodd = []
     for i in 1:n
         if i % 2 != 0
             append!(lodd,i)
         else
             append!(leven, i)
       end
     end
     println("\nodd number list : \t", lodd)
     println("even number list : \t", leven)

     enter a number:
     stdin> 20
     entered number : 20

     odd number list :       Any[1, 3, 5, 7, 9, 11, 13, 15, 17, 19]
     even number list :      Any[2, 4, 6, 8, 10, 12, 14, 16, 18, 20]
```

Programme 3.16 Even or odd?

converts a student's grade into a letter grade, but this code is built with if blocks and is given to show the difference between ***if*** and ***elseif*** code blocks.

Programme 3.19, since all conditions are met, the student's grade is classified according to all conditions. The difference between the operation of ***if-else*** and ***if-elseif*** code block can be clearly given in this way.

```
[5]: print("Please Enter the score of the student: ")
     n = parse(Int, readline(stdin))
     print("entered score : ", n)
     println()

     if n >= 89
         println("The student's score is AA")
     elseif n >= 85
         println("The student's score is BA")
     elseif n >= 80
         println("The student's score is BB")
     elseif n >= 75
         println("The student's score is BC")
     elseif n >= 70
         println("The student's score is CC")
     else n < 70
         println("The student is unsuccessful")
     end

     Please Enter the score of the student:
     stdin> 97
     entered score : 97
     The student's score is AA
```

Programme 3.17 A letter grade of a student's grade

```
[6]: print("Please Enter the score of the student: ")
     n = parse(Int, readline(stdin))
     print("entered score : ", n)
     println()

     if n >= 89
         println("The student's score is AA")
     elseif n >= 85
         println("The student's score is BA")
     elseif n >= 80
         println("The student's score is BB")
     elseif n >= 75
         println("The student's score is BC")
     elseif n >= 70
         println("The student's score is CC")
     else n < 70
         println("The student is unsuccessful")
     end

     Please Enter the score of the student:
     stdin> 68
     entered score : 68
     The student is unsuccessful
```

Programme 3.18 A letter grade of a student's grade

```
[2]: print("Please Enter the score of the student: ")
     n = parse(Int, readline(stdin))
     print("entered score : ", n)
     println()

     if n >= 89
         println("The student's score is AA")
         if n >= 85
             println("The student's score is BA")
         end
         if n >= 80
             println("The student's score is BB")
         end
         if n >= 75
             println("The student's score is BC")
         end
         if n >= 70
             println("The student's score is CC")
         end
         else n < 70
             println("The student is unsuccessful")
         end

     Please Enter the score of the student:
     stdin> 98
     entered score : 98
     The student's score is AA
     The student's score is BA
     The student's score is BB
     The student's score is BC
     The student's score is CC
```

Programme 3.19 A letter grade of a student's grade without ***elseif***

3.6 break and continue

In Julia, an infinite loop can be opened with while true. For example, if it is necessary to continuously receive data input from the keyboard, such an infinite loop can be preferred instead of running the programme repeatedly. However, in such a case, the infinite loop can be terminated with a control parameter. For this purpose, the ***break*** command is used, and the infinite loop is terminated when the condition is met. Programme 3.20 such an application is made. A number input is requested from the keyboard and each time it is asked whether the number entered is positive or negative. When the keyword "quit" is entered, the if conditional structure is activated and the infinite loop is exited with break and the programme is terminated.

The ***continue*** can be used to jump from this state to the next state when a certain condition is met. Programme 3.21 shows the use of the continue command so that the elements of the list are printed to the screen, but if the numbers are even numbers, then this number is skipped, and the next element is checked whether this element is an odd number or an even number.

```
[6]: while true
         print("enter a number: ")
         n = readline(stdin)

         if  n == "quit"
             println("The programme was over")
             break
         else
             n = parse(Int, n)
             if n % 2 == 0
                 println(string(n, " ", "is an even number"))
             else
                 println(string(n, " ", "is an odd number"))
             end
         end
     end

     enter a number:
     stdin>  12
     12 is an even number
     enter a number:
     stdin>  17
     17 is an odd number
     enter a number:
     stdin>  2024
     2024 is an even number
     enter a number:
     stdin>  quit
     The programme was over
```

Programme 3.20 Even or odd with an infinite loop

```
[7]: l = [i for i in 1:7]

     for i in l
         if i % 2 == 0
             continue
         else
             println(i)
         end
     end

     1
     3
     5
     7
```

Programme 3.21 The usage of *continue*

3.7 Functions in Julia

Defining functions in the Julia programming language is very flexible and there are several different ways to do it. The first of these methods is to define a function as in Python, but the ***function*** statement is used instead of ***def***, and the function is ended with ***end*** after the function is defined, as in loops. However, it should be noted that functions such as trigonometric functions, exponential functions, logarithms, etc., can be executed by calling them directly without loading any module, Programme 3.22.

The most basic of the methods that can be used to define functions in Julia is to define functions as if you were working on a worksheet. Programme 3.23 is a good example. The function $f(x) = x^2$ directly creates a function f and calculates the square

```
[1]: cos(deg2rad(45))
[1]: 0.7071067811865476
[2]: rad2deg(acos(0))
[2]: 90.0
[3]: log(exp(1))
[3]: 1.0
[4]: exp(-0.5)
[4]: 0.6065306597126334
[5]: abs(-5)
[5]: 5
[6]: log(exp(1),exp(2))
[6]: 2.0
[7]: sqrt(16)
[7]: 4.0
```

Programme 3.22 Standard defined functions in Julia

of the parameter value entered in this function. Similarly, many mathematical functions can be defined according to the purpose. In the second cell of Programme 3.23, a list is created by adding the values calculated using the f function defined in the previous cell to an empty list. Programme 3.24 represents another application for the function definition.

The second way to define a function in Julia is to use the ***function f(x)*** statement. In the code block opened with this statement, the operation to be performed by the function should be indented with the ***return*** statement in the code block and the code block should be terminated with ***end*** (Programmes 3.25, 3.26 and 3.27).

A third way to define a function in Julia is

```
function name = variable -> what the function will do
```

```
[8]: f(x) = x ^ 2
     f(4)
[8]: 16
[9]: L = []

     for i in 0:2:20
         append!(L, f(i))
     end
     print(L)
     Any[0, 4, 16, 36, 64, 100, 144, 196, 256, 324, 400]
```

Programme 3.23 The first function definition method

```
[10]: f(x) = sin(x)
      for i in 0:15:180
          println(i, "\t", round(f(deg2rad(i)), digits=8))
      end

      0       0.0
      15      0.25881905
      30      0.5
      45      0.70710678
      60      0.8660254
      75      0.96592583
      90      1.0
      105     0.96592583
      120     0.8660254
      135     0.70710678
      150     0.5
      165     0.25881905
      180     0.0
```

Programme 3.24 An application for the function definition

```
[11]: function g(x)
          u = 2*sin(x)
          return u*cos(x)
      end
      g(deg2rad(45))

[11]: 1.0
```

Programme 3.25 A second kind function definition in Julia

```
[12]: function f(n)
          if n == 0
              return 1
          else
              return n * f(n-1)
          end
      end

      n = 5
      println("the factorial of $n number is: ", f(n))

      the factorial of 5 number is: 120

[13]: n = 6
      println("the factorial of $n number is: ", f(n))

      the factorial of 6 number is: 720
```

Programme 3.26 A function for calculating the factorial of a number

This definition can be thought of as the Julia equivalent of lambda expressions in Python. It is a very practical way to define small functions. The three different function definition methods examined so far are important features of the flexible and fast structure of the Julia language.

Programme 3.28 shows a function of such an implementation. In the second cell of the programme, the calculation was made using the Euler formula. In Julia language, complex numbers are defined directly with the ***im*** term and without the need to load any module.

```
[14]: #= The following code is used for the recurrence relation
      for Legendre polynomials =#

      function P(l, x)
          if l == 0
              return 1
          elseif l == 1
              return x
          else
              return ((2*l-1)*x*P(l-1, x) - (l-1)*P(l-2, x)) / l
          end
      end

      x = 0.5
      println("P(3, ", x, ") = ", P(3, x))

      P(3, 0.5) = -0.4375

[15]: n = 5
      for i in 0:0.1:1
          println("x= ", i, "\t\t", "P($n, x)= ", round(P(n, i), digits = 9))
      end

      x= 0.0          P(5, x)= 0.0
      x= 0.1          P(5, x)= 0.17882875
      x= 0.2          P(5, x)= 0.30752
      x= 0.3          P(5, x)= 0.34538625
      x= 0.4          P(5, x)= 0.27064
      x= 0.5          P(5, x)= 0.08984375
      x= 0.6          P(5, x)= -0.15264
      x= 0.7          P(5, x)= -0.36519875
      x= 0.8          P(5, x)= -0.39952
      x= 0.9          P(5, x)= -0.04114125
      x= 1.0          P(5, x)= 1.0
```

Programme 3.27 The recurrence relation for the first kind of Legendre polynomials

The function structures analysed so far are given depending on a single variable. It is possible to define functions in more than one variable. Below is an example of ordinary function definitions defined for more than one variable (Programme 3.29).

3.8 Some Built-in Functions in Julia

In order to avoid the need to write separate code for some data types, the Julia programming language includes some built-in functions. In the following, these built-in functions will be briefly mentioned, and examples will be made on an application.

3.8.1 map()

It is clearly seen that the operation performed in 5 lines in the first cell of Programme 3.30 is performed in the first 2 lines of the second cell of the same programme. The map, which is a built-in function in Julia, could be used over an iterable function as in the programme.

```
[16]: csc = x -> 1/sin(x)

      for i in 0:30:180
          csc(i)
          println("i: ", i, "\t", csc(deg2rad(i)))
      end

      i: 0    Inf
      i: 30   2.0000000000000004
      i: 60   1.1547005383792517
      i: 90   1.0
      i: 120  1.1547005383792515
      i: 150  2.0000000000000004
      i: 180  8.165619676597685e15

[17]: f1= t -> cos(t) + im * sin(t)

      for i in 0:30:180
          f1(i)
          println("i: ", i, "\t\t", "f1($i) = ", round(f1(deg2rad(i)),digits=8))
      end

      i: 0            f1(0) = 1.0 + 0.0im
      i: 30           f1(30) = 0.8660254 + 0.5im
      i: 60           f1(60) = 0.5 + 0.8660254im
      i: 90           f1(90) = 0.0 + 1.0im
      i: 120          f1(120) = -0.5 + 0.8660254im
      i: 150          f1(150) = -0.8660254 + 0.5im
      i: 180          f1(180) = -1.0 + 0.0im
```

Programme 3.28 A third way in order to define a function in Julia

3.8.2 *reduce()*

Another built-in function is the ***reduce()***. It applies the function, which is definite before, to the iterable data with the related argument. The working principle of the ***reduce()*** function is given in Fig. 3.1. Summation process is applied to the first two elements of the list, then the result is applied to the third element of the list, the result is applied to the fourth element of the list, and so on.

Programme 3.31, two operations were performed. In the first operation, the sum of the list elements was taken and in the second operation, the list elements were multiplied. Figure 3.1 shows the process of summation. Programme 3.31, it is sufficient to write only the multiplication symbol (*) or the addition symbol (+). The same operation in Python requires a lambda expression.

The reduce function can also be used to find the maximum or minimum item in a list or tuple data type, as in cells 6 and 7 in Programme 3.31, cells 6 and 7 can also be used to find the maximum or minimum element in a list or tuple data type. In fact, in Programme 3.31 as ***reduce(max, L)*** and ***reduce(min, L)*** in cells six and seven, respectively, in Programme 3.32 to determine both the maximum and minimum element.

```
[18]: f(x, y) = log(x) * exp(y)
      f(2,1)

[18]: 1.88416938536372

[19]: v(a, t) = vo + a * t

      vo, a = 0, 3
      for j in 0:0.5:2
          println("a= ", a, "\t", "t= ", j, "\t", "v($a,t)= ", v(a, j))
      end

      a= 3   t= 0.0  v(3,t)= 0.0
      a= 3   t= 0.5  v(3,t)= 1.5
      a= 3   t= 1.0  v(3,t)= 3.0
      a= 3   t= 1.5  v(3,t)= 4.5
      a= 3   t= 2.0  v(3,t)= 6.0

[20]: function v(a, t)
          vo = 2
          return vo + a * t
      end

      for a in 0:1:2
          for t in 0:0.5:2
              println("a= ", a, "\t", "t= ", t, "\t", "v($a,t)= ", v(a, t))
          end
      end

      a= 0   t= 0.0  v(0,t)= 2.0
      a= 0   t= 0.5  v(0,t)= 2.0
      a= 0   t= 1.0  v(0,t)= 2.0
      a= 0   t= 1.5  v(0,t)= 2.0
      a= 0   t= 2.0  v(0,t)= 2.0
      a= 1   t= 0.0  v(1,t)= 2.0
      a= 1   t= 0.5  v(1,t)= 2.5
      a= 1   t= 1.0  v(1,t)= 3.0
      a= 1   t= 1.5  v(1,t)= 3.5
      a= 1   t= 2.0  v(1,t)= 4.0
      a= 2   t= 0.0  v(2,t)= 2.0
      a= 2   t= 0.5  v(2,t)= 3.0
      a= 2   t= 1.0  v(2,t)= 4.0
      a= 2   t= 1.5  v(2,t)= 5.0
      a= 2   t= 2.0  v(2,t)= 6.0
```

Programme 3.29 Some simple functions defined with more than one variable in Julia

```
[1]: f(x) = x ^ 2

     L = []

     for i in 0:2:20
         append!(L, f(i))
     end
     print(L)

     Any[0, 4, 16, 36, 64, 100, 144, 196, 256, 324, 400]

[2]: g(x) = x ^ 2
     M = map(g, 0:2:20)
     print(M)

     [0, 4, 16, 36, 64, 100, 144, 196, 256, 324, 400]

[3]: length(M)

[3]: 11

[4]: sqrt(M[length(M)])

[4]: 20.0
```

Programme 3.30 The usage of *map()* function, which is a built-in function

```
[5]: L = [1/2, 3/2, 4/2, 5/2, 6/2]

     sum = reduce(+, L)
     println(sum)

     pro =reduce(*, L)
     println(pro)

     9.5
     11.25

[6]: reduce(max, L)

[6]: 3.0

[7]: reduce(min, L)

[7]: 0.5
```

Programme 3.31 The application of *reduce()* function

Fig. 3.1 The operation of the *reduce()* function

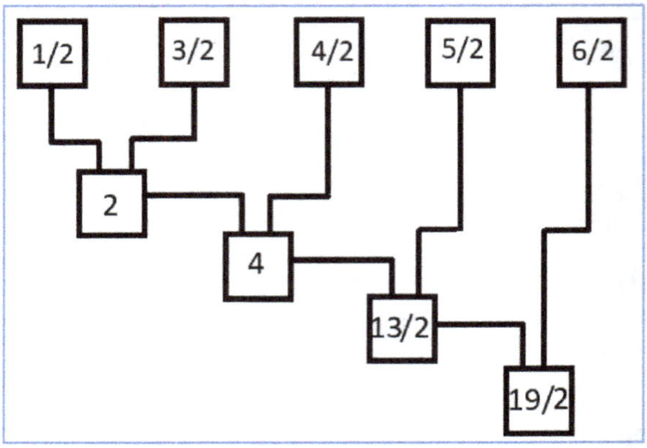

3.8.3 *filter()*

The ***filter()*** function takes two parameters, the first parameter is a logical function, and the second parameter is an iterable data type such as list or tuple. Programme 3.22 checks for an even number.

Programme 3.33 checks whether a number is a prime number or not and determines the prime numbers by checking whether the numbers in the given range are prime or not and prints them on the screen (Programme 3.34).

3.8.4 *zip() and collect() functions*

This function concatenates index by index list elements and returns them as tuple data type. For example, Programme 3.35, two list types of equal length are concatenated

```
[8]: function maks(x, y)
         if x > y
             return x
         else
             return y
         end
     end
     reduce(maks,L)

[8]: 3.0

[9]: function mini(x, y)
         if x < y
             return x
         else
             return y
         end
     end
     reduce(mini,L)

[9]: 0.5
```

Programme 3.32 Maximum and minimum functions without *reduce()* function

```
[10]: even(n) = n % 2 == 0
      c = filter(even, 0:10)
      println(c)

      [0, 2, 4, 6, 8, 10]
```

Programme 3.33 Even or odd number?

```
[11]: function pri(x)
          i = 2
          if x == 1
              return false
          elseif x == 2
              return true
          else
              while i < x
                  if x % i == 0
                      return false
                  end
                  i += 1
              end
              return true
          end
      end

[11]: pri (generic function with 1 method)

[12]: L = [filter(pri,1:50)]
      L

[12]: 1-element Vector{Vector{Int64}}:
       [2, 3, 5, 7, 11, 13, 17, 19, 23, 29, 31, 37, 41, 43, 47]
```

Programme 3.34 Prime numbers with *filter()* function

with the ***zip()*** function. This function is useful for traversing two lists at the same time. When the ***zip()*** function is applied to the dictionary data type, the keys match.

Programme 3.36, the for loop is applied to the data *z* obtained by applying the zip function to two list data types. Although it is not clearly visible in the output of the

```
[13]: l1 = [1, 2, 3, 4, 5]
      l2 = ["a", "b", "c", "d", "e"]

      z = zip(l1,l2)

[13]: zip([1, 2, 3, 4, 5], ["a", "b", "c", "d", "e"])

[14]: first(z)

[14]: (1, "a")

[15]: length(z)

[15]: 5
```

Programme 3.35 The application of *zip()* function

```
[16]: for z0 in z
          println(z0)
      end

      (1, "a")
      (2, "b")
      (3, "c")
      (4, "d")
      (5, "e")

[17]: for (x, y) in z
          println(x, ", ", y)
      end

      1, a
      2, b
      3, c
      4, d
      5, e

[18]: coll = collect(z)
      println(coll)

      [(1, "a"), (2, "b"), (3, "c"), (4, "d"), (5, "e")]
```

Programme 3.36 The elements of the zipped data and the usage of the *collect()* function

13th cell, when printed with for loop, the elements of the lists ***l1*** and ***l2*** are combined index by index. Without using the for loop, the ***collect ()*** function returns a list data type consisting of the zip elements. The elements of this list are the tuples formed by concatenating the elements from the lists ***l1*** and ***l2***.

3.8.5 *enumerate() function*

This function is used to display iterable data set elements such as a list or tuple with their index numbers using a for loop (Programme 3.37).

```
[20]: for (index, value) in enumerate(z)
          println("$index $value")
      end

      1 (1, "a")
      2 (2, "b")
      3 (3, "c")
      4 (4, "d")
      5 (5, "e")

[21]: L =["electron", "proton", "neutron"]

      for (index, value) in enumerate(L)
          println("$index $value")
      end

      1 electron
      2 proton
      3 neutron
```

Programme 3.37 The usage of the *enumerate()* function

3.8.6 all() and any() functions

The ***all()*** function checks whether all elements in the iterable data are correct. It takes a logical function as the first parameter and an iterable data set as the second parameter. If the result of the logical function is true for all elements, it returns ***true***, otherwise it returns ***false***. It can be thought of as the application of the and statement in logic. ***any()*** function works like the application of the or statement in logic. If there is at least one condition that satisfies the applied logical function, the result returns ***true***, otherwise it returns ***false*** (Programme 3.38).

```
[23]: L = [i for i in 1:20]
      all(x-> x % 2 == 0, L)

[23]: false

[24]: L = [i for i in 0:2:20]
      all(x-> x % 2 == 0, L)

[24]: true

[25]: L = [i for i in 1:20]
      any(x-> x % 2 == 0, L)

[25]: true

[26]: L = [i for i in 1:2:20]
      any(x-> x % 2 == 0, L)

[26]: false
```

Programme 3.38 The usage of *all()* and *any()* functions

3.9 Studying in Julia with Special Functions in Physics

Special Functions have a great importance in Physics and Engineering. In electromagnetic theory, Bessel functions are encountered when the propagation and scattering of electromagnetic waves are analysed in problems with cylindrical symmetry. The solutions of the Schrödinger wave equation of the hydrogen atom are expressed by Laguerre polynomials for the spatial part and spherical harmonics for the angular part. The most important properties of these special functions are that they are orthogonal to the geometry in which they are defined and can be expressed in terms of higher order terms by means of recurrence relations.

Using the loop structures analysed above, these recurrence relations can be written by any user. For example, Programme 3.26 gives the recurrence relation for Legendre polynomials. Similarly, recurrence relations can be written for other special functions. However, special modules have been prepared in Julia language. In this section we will give brief information about these modules.

3.9.1 *Legendre Polynomials*

The *LegendrePolynomails* module [2] is used in Julia to compute the value of Legendre polynomials and associated Legendre polynomials. To install the package, the Julia terminal must first be loaded with the package loader command line and then the corresponding package must be loaded by calling this package loader, see Programme 3.39.

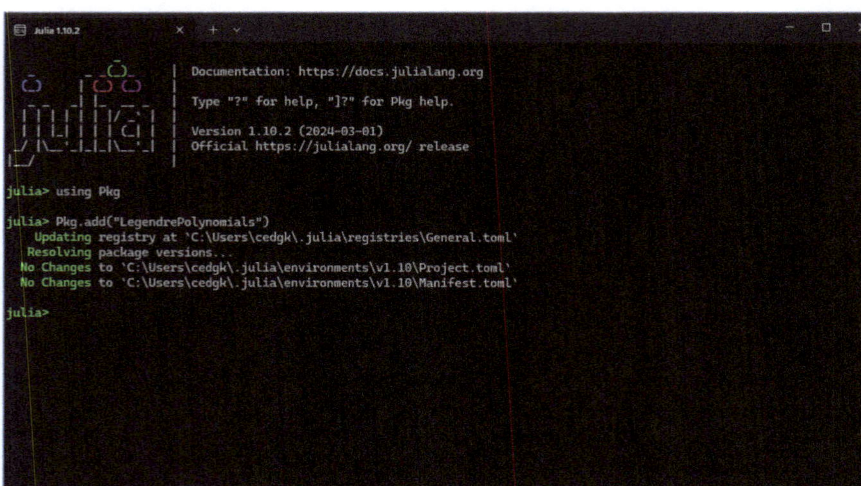

Programme 3.39 The process of installing Legendre polynomials

In the first cell in Programme 3.40, the *LegendrePolynomials* package is included in the programme. In the second cell, the value of the second order first type Legendre polynomial for $x = 0.1$ is calculated. In the third cell, the value of the Legendre polynomials from zero to seventh order for $x = 0.1$ is printed with the ***collectPL()*** function. In the fourth cell of the programme, the value of the first derivative of the first derivative of the second order Legendre polynomial for $x = 0.1$ is printed. In the cell immediately after it, the value of the second derivative of the second order first type Legendre polynomial at $x = 0.1$ is calculated. In the sixth cell, the function ***collecdnPl()*** is used to print the value of the third derivative of the first 7 Legendre polynomials at the point 0.1.

collectPl() function takes two parameters. The first parameter specifies the point at which the value of the function is to be calculated and the upper order value to be calculated with = ***(any integer)***. The function ***collectdnPl()*** takes three parameters. The first parameter specifies the point where the derivative is to be calculated, the second parameter specifies the order value with ***lmax = (any integer)*** and the order of the derivative with ***n = (any integer)***.

The same module can be used to graph Legendre polynomials. ***Plots*** module [4] should be used for this purpose. A comprehensive review and details of this module

```
[1]: using LegendrePolynomials

[2]: Pl(0.1,2)

[2]: -0.485

[3]: collectPl(0.1; lmax = 7)

[3]: 8-element OffsetArray(::Vector{Float64}, 0:7) with eltype Float64 with indices 0:7:
       1.0
       0.1
      -0.485
      -0.1475
       0.3379375
       0.17882875
      -0.24882931249999998
      -0.19949294375

[4]: dnPl(0.1,2,1) # first derivation of second order Legendre polynom

[4]: 0.30000000000000004

[5]: dnPl(0.1,2,2)

[5]: 3.0

[6]: collectdnPl(0.1, lmax=7, n=3)

[6]: 8-element OffsetArray(::Vector{Float64}, 0:7) with eltype Float64 with indices 0:7:
        0.0
        0.0
        0.0
       15.0
       10.500000000000002
      -47.775
      -45.51750000000001
       92.7005625
```

Programme 3.40 Some operations with Legendre polynomials

```
[7]: using Plots

     l = 5
     x = -1:0.01:1
     y = [Pl(val, l) for val in x]

     plot(x, y, label="P_$(l)(x)")
```

[7]:

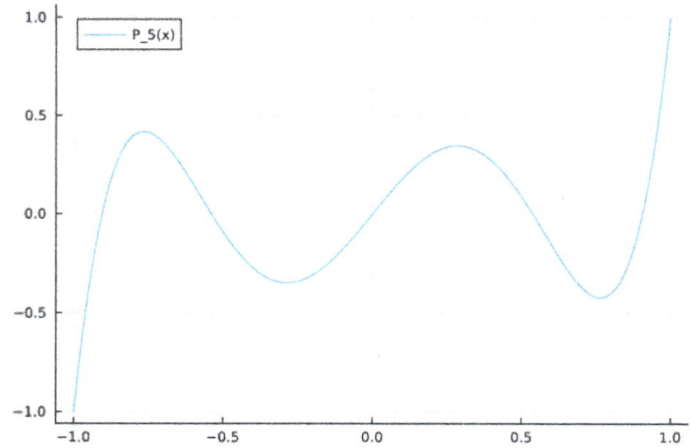

Programme 3.41 The graphic of P(val, 5) with the Plots module

will be analysed in the next section. In this section, it is reasonable to mention the graphing of special functions.

In Programme 3.41, the graph of the first type of Legendre polynomial of the fifth order is plotted in the interval $[-1, 1]$, which is known to be the range of Legendre polynomials. Programme 3.42 represents the graphics of first five Legendre polynomials. Similarly, the derivatives of Legendre polynomials can be plotted as in Programme 3.43.

Programme 3.44 represents the calculation of the values of $P_2^1(0.1)$ and $P_3^2(0.5)$, respectively. These are associated Legendre polynomials. Programme 3.45 represents the graphics of the associated Legendre polynomials for $\ell = 2, 3, 4, 5$ and $m = 1.$. For a comprehensive review of Legendre polynomials, see Bell (1968).

3.9.2 *Bessel Functions with Bessels Module*

Julia has two modules for working with Bessel functions. The first one is the *Bessels* module [5]. This module can be used for Bessel functions, modified Bessel functions, Hankel functions, spherical Bessel functions and both the first and second kind of Airy functions. The other one is the *SpecialFunctions* module. This module can be used for Bessel, Hankel, Airy, error function (erf), eta, zeta, etc.

Programme 3.46 shows the numerical results of some Bessel functions after loading the *Bessels* module. The first parameter is the parameter n and the second

```
[8]: x = -1:0.01:1

     p = plot()

     for l in 0:4
         y = [Pl(val, l) for val in x]
         plot!(p, x, y, label="P_$(l)(x)")
     end

     display(p)
```

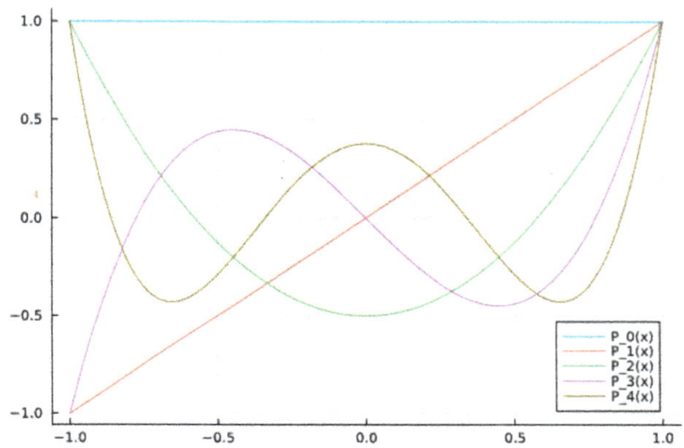

Programme 3.42 The graphics of first five Legendre polynomials

```
[9]: x = -1:0.01:1

     p = plot()

     for l in 0:4
         y = [dnPl(val, l, 1) for val in x]
         plot!(p, x, y, label="P_$(l)(x)")
     end

     display(p)
```

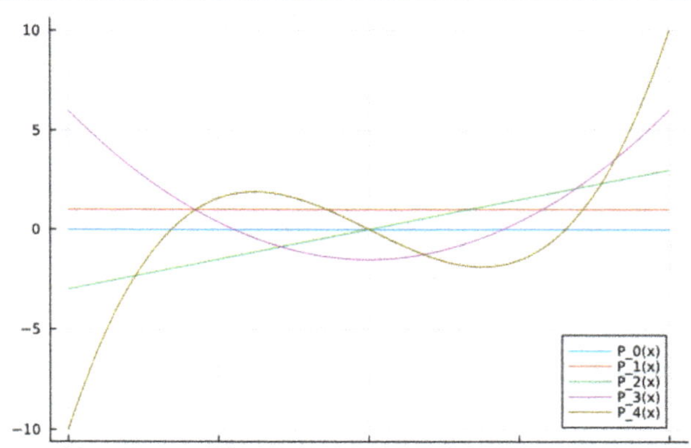

Programme 3.43 Graph of the first derivatives of the first five Legendre polynomials

```
[10]: Plm(0.1, 2, 1)

[10]: -0.29849623113198603

[11]: l, m, x = 3, 2, 0.5
      p_lm = Plm(x, l, m)

[11]: 5.624999999999997
```

Programme 3.44 The values of $P_2^1(0.1)$ and $P_3^2(0.5)$

```
[12]: x_vals = -1:0.01:1
      m = 1
      plot_legendre = plot()

      for l in 2:5
          y_vals = [Plm(x, l, m) for x in x_vals]
          plot!(plot_legendre, x_vals, y_vals, label="P_$(l)^$m(x)")
      end

      display(plot_legendre)
```

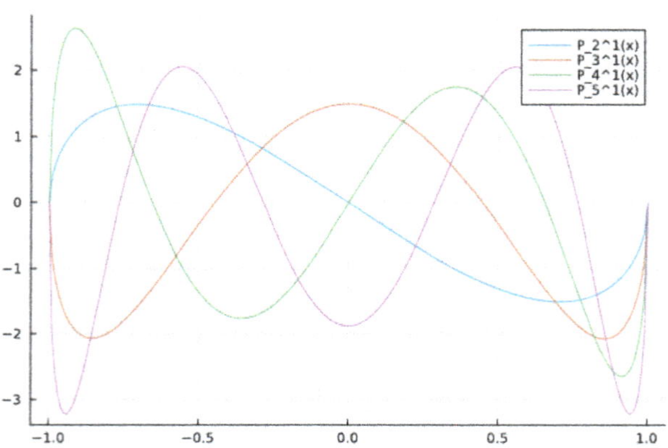

Programme 3.45 The graphics of the associated Legendre polynomials for $\ell = 2, 3, 4, 5$ and $m = 1$

parameter is the argument. Programme 3.47 calculates the values of the Bessel functions of the first kind for the first five Bessel function from $v = 0$ to $v = 4$ for the range of $\boldsymbol{x}$ values from 0 to 20 and plot them on the same graph (Table 3.1 and Programme 3.48).

3.9.3 *Bessel Functions with SpecialFunctions Module*

SpecialFunctions [6] module contains a rich content of special functions which based on the OpenSpecFun [7] and OpenLibm [8] libraries. Table 3.2 shows the functions that can be used with this module (Programmes 3.49 and 3.50).

```
[13]: using Bessels

[14]: besselj(0, 5.2) # (first parameter n and second parameter z)

[14]: -0.11029043979098647

[15]: besselj(0, 4)

[15]: -0.3971498098638474

[16]: l = [besselj(0.5, z) for z in 0:4]

[16]: 5-element Vector{Float64}:
        0.0
        0.6713967071418033
        0.5130161365618278
        0.06500818287737574
       -0.3019205132916391

[17]: ll = [besselj((2*i+1)/2, 0.1) for i in 0:2]

[17]: 3-element Vector{Float64}:
       0.2518929403260009
       0.008402034301500145
       0.0001680887190033413
```

Programme 3.46 The values of some Bessel function with *Bessels* module

```
[25]: x_vals = 0:0.1:20

      p = plot()

      for v in 0:4
          y_vals = [besselj(v, x) for x in x_vals]
          plot!(p, x_vals, y_vals, label="J_$(v)(x)")
      end

      display(p)
```

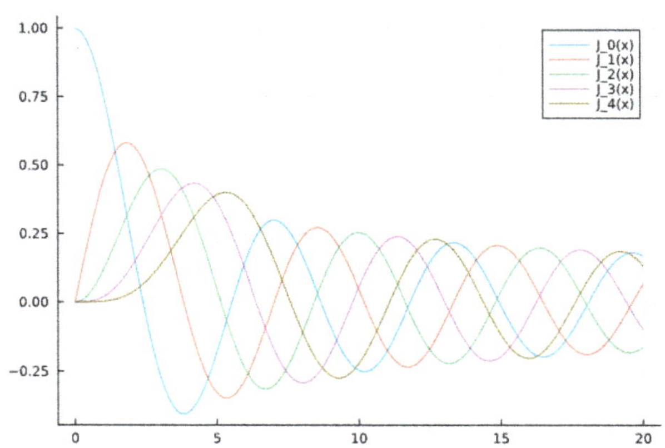

Programme 3.47 The graphics for first five Bessel functions

Table 3.1 Bessel functions that can be used with *Bessels* module

Function	Function
`besselj0(x)`	`besselh(nu, k, x)`
`besselj1(x)`	`hankelh1(nu, x)`
`besselj(nu, x)`	`hankelh2(nu, x)`
`bessely0(x)`	`airyai(x)`
`bessely1(x)`	`airyaiprime(x)`
`bessely(nu, x)`	`airybi(x)`
`besseli0(x)`	`airybiprime(x)`
`besseli1(x)`	`sphericalbesselj(nu, x)`
`besseli(nu, x)`	`sphericalbessely(nu, x)`
`besselk0(x)`	`Bessels.sphericalbesseli(nu, x)`
`besselk1(x)`	`Bessels.sphericalbesselk(nu, x)`
`besselk(nu, x)`	`Bessels.gamma(x)`

```
[28]: x_vals = 0:0.1:5

      p = plot()

      for v in 0:4
          y_vals = [besseli(v, x) for x in x_vals]
          plot!(p, x_vals, y_vals, label="J_$(v)(x)")
      end

      display(p)
```

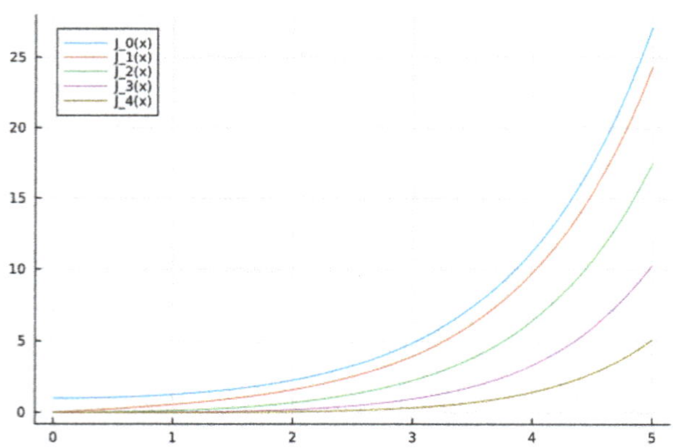

Programme 3.48 The graphics for first five *besseli(nu, x)* functions

3.10 Macros in Julia

In 1958, John McCarthy developed the Lisp language, the main feature of which is the ability to perform operations on lists, and for this reason it is known by the abbreviation of List Processing. Lisp programming language includes functional, procedural and meta programming features. With this feature, it is used in artificial intelligence and symbolic calculations.

The following text was taken from Ref. [1] which is Julia documentation web source.

Table 3.2 The functions that can be used with *SpecialFunctions* module

Function	Description
`erf(x)`	Error function at x
`erfc(x)`	Complementary error function, i.e. the accurate version of 1-erf(x) for large x
`erfinv(x)`	Inverse function to erf()
`erfcinv(x)`	Inverse function to erfc()
`erfi(x)`	Imaginary error function defined as -im * erf(x * im), where im is the imaginary
`erfcx(x)`	Scaled complementary error function, i.e. accurate exp(x^2) * erfc(x) for large x
`dawson(x)`	Scaled imaginary error function, a.k.a. Dawson function
`digamma(x)`	Digamma function (i.e. the derivative of lgamma at x)
`eta(x)`	Dirichlet eta function at x
`zeta(x)`	Riemann zeta function at x
`airyai(z)`	Airy Ai function at z
`airyaiprime(z)`	Derivative of the Airy Ai function at z
`airybi(z)`	Airy Bi function at z
`airybiprime(z)`	Derivative of the Airy Bi function at z
`airyaix(z), airyaiprimex(z), airybix(z), airybiprimex(z)`	Scaled Airy Ai function and kth derivatives at z
`besselj(nu,z)`	Bessel function of the first kind of order nu at z
`besselj0(z)`	besselj(0,z)
`besselj1(z)`	besselj(1,z)
`besseljx(nu,z)`	Scaled Bessel function of the first kind of order nu at z
`bessely(nu,z)`	Bessel function of the second kind of order nu at z
`bessely0(z)`	bessely(0,z)
`bessely1(z)`	bessely(1,z)
`besselyx(nu,z)`	Scaled Bessel function of the second kind of order nu at z
`besselh(nu,k,z)`	Bessel function of the third kind of order nu at z; k must be either 1 or 2
`hankelh1(nu,z)`	besselh(nu, 1, z)
`hankelh1x(nu,z)`	Scaled besselh(nu, 1, z)
`hankelh2(nu,z)`	besselh(nu, 2, z)
`hankelh2x(nu,z)`	Scaled besselh(nu, 2, z)
`besseli(nu,z)`	Modified Bessel function of the first kind of order nu at z
`besselix(nu,z)`	Scaled modified Bessel function of the first kind of order nu at z
`besselk(nu,z)`	Modified Bessel function of the second kind of order nu at z
`besselkx(nu,z)`	Scaled modified Bessel function of the second kind of order nu at z

* This table is taken from [6]

```
[38]: using SpecialFunctions

[39]: erf(0.1)

[39]: 0.1124629160182849

[40]: erfinv(0.1)

[40]: 0.08885599049425769

[41]: besselj0(0.1)

[41]: 0.99750156206604

[42]: zeta(0.3)

[42]: -0.9045592572539837
```

Programme 3.49 The usage of some functions in *SpecialFunctions* module

```
[43]: x_vals = 0:0.1:10
      p = plot()
      for v in 0:4
          y_vals = [besselj(v, x) for x in x_vals]
          plot!(p, x_vals, y_vals, label="J_$(v)(x)")
      end
      display(p)
```

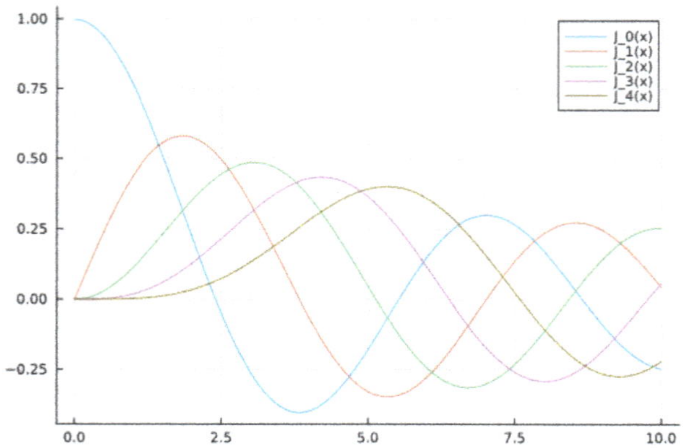

Programme 3.50 The graphics for first five Bessel functions with *SpecialFunctions* module

The Lisp programming language has a great influence on the design of the Julia programming language and especially the macro feature is also seen in the Julia language. For this reason, Julia can be referred to as a language that carries the legacy of the Lisp language. The Julia programming language combines the functional programming and macro advantages of the Lisp language with modern programming language and performance.

The intended use of macros can be listed as follows:

Code Generation: Macros can automatically generate code that repeats certain patterns. The use of macros can optimise the process when a similar function is to be defined.
Performance Optimisation: In some cases, macros can improve performance by performing calculations at compile time rather than at run time.
Syntax Simplifications: Macros extend the syntax of the language, making it possible to write more readable or expressive code.
Error Control: Macros can be used for error checking and management.

The use of macros is powerful but can lead to complexity and make the code difficult to understand. For this reason, the Julia documentation recommends that they should be used when higher-order functions cannot be implemented.

To define a macro in Julia language @ symbol is used. Programme 3.51 represents an example to use a macro. In the created sinus macro, the ***esc*** function is used to use variables correctly. The use of ***eval*** and ***Meta.parse*** allows the macro to be evaluated for each element.

```
[1]: macro sinus(x)
         esc(:(sin(deg2rad($x))))
     end

[1]: @sinus (macro with 1 method)

[2]: @sinus 30

[2]: 0.49999999999999994

[3]: angles = [0, 30, 45, 60, 90]
     sin_values = map(x -> eval(Meta.parse("@sinus $x")), angles)

[3]: 5-element Vector{Float64}:
      0.0
      0.49999999999999994
      0.7071067811865475
      0.8660254037844386
      1.0

[4]: for i in angles
         println(@sinus i)
     end

     0.0
     0.49999999999999994
     0.7071067811865475
     0.8660254037844386
     1.0
```

Programme 3.51 *Macro* definition in Julia and usage of it in map function

Programme 3.52, the else block is entered to give the error message "Unsupported function" when a different character string or a different function name is entered from the return line, which is one of the purposes of macros, Programme 3.53.

```
[5]: macro trigfunc(func, x)
         if func in [:sin, :cos, :tan]
             return esc(:( $(Symbol(func))(deg2rad($x)) ))
         else
             throw(ArgumentError("Unsupported function"))
         end
     end

[5]: @trigfunc (macro with 1 method)

[6]: @trigfunc sin 30

[6]: 0.49999999999999994

[7]: @trigfunc cos 45

[7]: 0.7071067811865476
```

Programme 3.52 An effective *Macro* definition for trigonometric functions

```
[8]: @trigfunc cs 45

     LoadError: ArgumentError: Unsupported function
     in expression starting at In[8]:1

     Stacktrace:
      [1] var"@trigfunc"(__source__::LineNumberNode, __module__::Module, func::Any, x::Any)
        @ Main .\In[5]:5
```

Programme 3.53 Error message with macro

```
[9]:  using Bessels

[10]: macro besselfunc(n, x)
          esc(:(besselj($n, $x)))
      end

[10]: @besselfunc (macro with 1 method)

[11]: @besselfunc 0 1.0

[11]: 0.7651976865579666

[12]: @besselfunc 1 1.0

[12]: 0.44005058574493355

[13]: using LegendrePolynomials

[14]: macro legendrefunc(n, x)
          esc(:(Pl($x, $n)))
      end

[14]: @legendrefunc (macro with 1 method)

[15]: @legendrefunc 2 0.5

[15]: -0.125
```

Programme 3.54 *Besselj* and Legendre polynomials with macro structure

Programme 3.53 is a macro for the Bessel function. Thus, as shown in the same programme, the value of the Bessel function can be calculated for the desired value of n and at the desired point x in a much more practical way. Programme 3.54 is a macro for Legendre polynomials.

3.11 do structure in Julia

This structure can be used to create anonymous or lambda functions and to use them in a specific context. It is especially useful when one function is an argument to another function. Therefore, it can often be used in iteration, data manipulation or to define the last argument of a function. ***do*** structure also helps to write more understandable code.

In Programme 3.55, a list of numbers from 0 to 20 is created and using the ***filter()*** function, even numbers in this list are filtered as *L_even* and *L_odd* in the next cell as lists of even and odd numbers. The same operations were used in the 4th and 5th cells of the programme with do structure.

In Programme 3.56, the *LegendrePolynomials* module and the do structure are used with the ***map()*** function to calculate the values of it for the second order Legendre polynomial and the graph of the function is drawn according to the values of ***x*** and the polynomial.

In the next example, the first 5 Legendre polynomials are plotted using the ***map()*** function and *do* structure. As can be seen, the do structure provides a flexible structure for iterative operations (Programme 3.57).

```
[1]: L = [i for i in 0:20]
     println(L)

     [0, 1, 2, 3, 4, 5, 6, 7, 8, 9, 10, 11, 12, 13, 14, 15, 16, 17, 18, 19, 20]

[2]: L_even = filter(i -> i % 2 == 0, L)
     println(L_even)

     [0, 2, 4, 6, 8, 10, 12, 14, 16, 18, 20]

[3]: L_odd = filter(i -> !(i % 2 == 0), L)
     println(L_odd)

     [1, 3, 5, 7, 9, 11, 13, 15, 17, 19]

[4]: Lis_even = filter(L) do i
         i % 2 == 0
     end
     println(Lis_even)

     [0, 2, 4, 6, 8, 10, 12, 14, 16, 18, 20]

[5]: Lis_odd = filter(L) do i
         !(i % 2 == 0)
     end
     println(Lis_odd)

     [1, 3, 5, 7, 9, 11, 13, 15, 17, 19]
```

Programme 3.55 Do code block and its application

```
[7]: using LegendrePolynomials

     x_values = 0:0.05:1

     Pl2_values = map(x_values) do x
         Pl(x, 2)
     end

     println("x values: ", x_values)
     println("Pl_2(x) values: ", Pl2_values)

     x values: 0.0:0.05:1.0
     Pl_2(x) values: [-0.5, -0.49625, -0.485, -0.46625, -0.44, -0.40625, -0.365, -0.31625000000000003, -0.25999
     999999999995, -0.19624999999999998, -0.125, -0.04624999999999999, 0.039999999999999925, 0.1337500000000000
     4, 0.23499999999999988, 0.34375, 0.4600000000000002, 0.58375, 0.7150000000000001, 0.8537499999999998, 1.0]

[8]: using Plots

     plot(x_values, Pl2_values, label = "P_2(x)", size = (400, 300))

[8]:
```

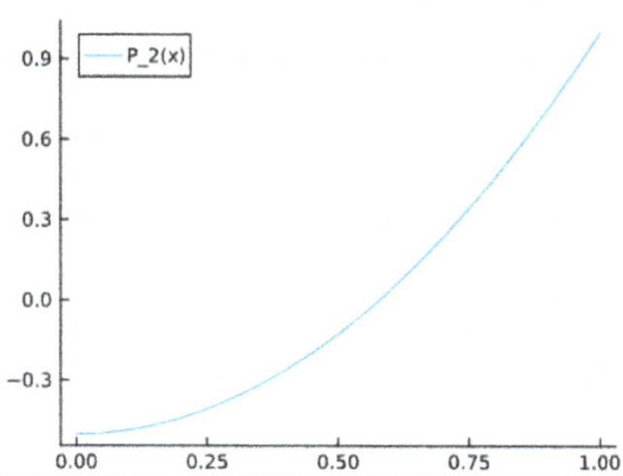

Programme 3.56 Legendre Polynomial with do structure

```
[9]: using LegendrePolynomials, Plots

x_values = -1:0.01:1

plt = plot(title="First 5 Legendre Polynomials",
    xlabel="x", ylabel="Pl(x, n)", legend=:bottomright)

for n in 0:5
    plot!(x_values, map(x_values) do x
        Pl(x, n)
    end, label="Pl_$n(x)")
end

display(plt)
```

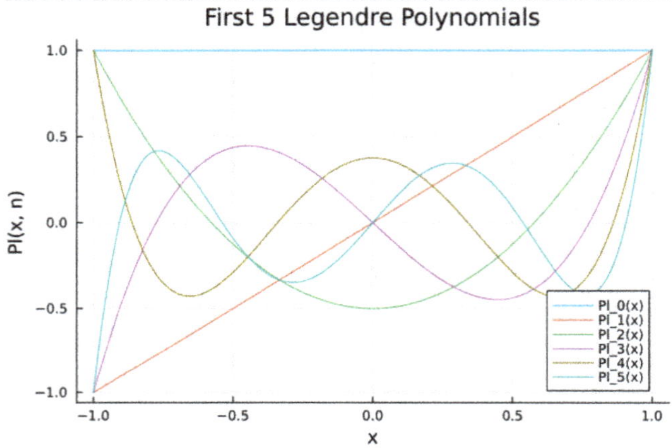

Programme 3.57 The graphics first 5 Legendre polynomials with *map()* function and do structure

3.12 Problems

1. Write a code to be able to the recurrence relation for second kind Legendre polynomials.
2. In the section on special functions, there is no explanation for the calculation of the second kind Legendre polynomials. Write a code to calculate and to plot the graph of the second kind of Legendre polynomials with the help of a source such as Ref. [4].
3. Write a code to be able to calculate the Hydrogen spectrum.
4. The Gamma function is given as

$$\Gamma(x) = \int_0^\infty e^{-t} t^{x-1} dx.$$

$\Gamma(1) = 1$ and if x is an integer, then $\Gamma(x+1) = x\Gamma(x) = x!$. If x is ½, then $\Gamma(1/2) = \sqrt{\pi}$. Write a code that calculates the value of the Gamma function for the cases where x is a whole number or a half-integer. Compare your results with the results of the gamma function available in the *SpecialFunctions* module.

5. Write a code to calculate the first 20 terms of Fibonacci sequence.
6. Write three codes that calculate the sum of odd numbers from 1 to 1000, differently even numbers and differently prime numbers.
7. Write a code that saves the elements of the Fibonacci array up to the entered value in a list by taking data input from the keyboard and prints the elements of this array on the screen according to the index number.
8. Write a code for the Taylor expansion of the logarithm function and determine to which term of the Taylor expansion the 16th digit of this expansion should be expanded in order to converge to the real value of the logarithm function.
9. Consider the seventh question for the exponential function, sine and cosine functions.
10. Write a function to calculate the Riemann integral of the sine function in the interval [0, pi/2] and calculate the value of the integral. Do this for 100, 1000 and 10000 divisions and see how different the result is for each of the three cases.

References

1. Julia Programming Language. https://julialang.org/; MIT License. 2011–2019: Jeff Bezanson, Stefan Karpinski, Viral B. Shah, and other contributors: https://github.com/JuliaLang/www.julialang.org/contributors
2. Legendre Polynomials module. https://juliapackages.com/p/legendrepolynomials (MIT License)
3. Plots module. https://docs.juliaplots.org/stable/tutorial/#tutorial (MIT License)
4. William W. 1968. Bell Special Functions for Scientists and Engineers. Dover Publications. ISBN 0-486-43521-0
5. Bessels module. https://juliapackages.com/p/bessels MIT License. 2021–2023, Michael Helton, Oscar Smith, Chris Geoga, and contributors. https://juliamath.github.io/Bessels.jl/stable/contribute/#Contributors-Guide-1
6. SpecialFunctions module. https://specialfunctions.juliamath.org/v0.1/ MIT License. 2017 Jeff Bezanson, Stefan Karpinski, Viral B. Shah, and others: https://github.com/JuliaMath/SpecialFunctions.jl/graphs/contributors
7. OpenSpecFun. https://github.com/JuliaMath/openspecfun; MIT License. 2009–2024, OpenSpecFun contributors. https://github.com/JuliaMath/openspecfun/contributors
8. OpenLibm. https://github.com/JuliaMath/openlibm; MIT License, 2011-14 The Julia Project. https://github.com/JuliaMath/openlibm/graphs/contributors

Chapter 4
Plots in Julia

4.1 Introduction

Plots and diagrams play a role in scientific communication equal in importance to words and equations. For this reason, we have denoted Julia's applications to process data and effectively visualise the results as plots. The variety of plots and their easy definitions are one of the key features of Julia by an external package. Hence, this chapter has also demonstrated how to work with external packages and defined their configurations.

The plotting ecosystem of Julia is diverse and powerful including the following list of packages:

PlotlyJS
Pyplots
Gr
Unicode Plots
Plotly
Gadify
Bokeh
Plots.

Plots is a visualization interface and toolset. It sits above other backends, like GR, PythonPlot, PGFPlotsX or Plotly connecting commands with implementation. If one backend does not support your wanted features, you can just switch to another backend with one command. It is not necessary to change your code and to learn a new syntax.

The goals with the package are [1]:

- **Powerful**. Do more with less. Complex visualizations become easy.
- **Intuitive**. Start generating plots without reading volumes of documentation. Commands should "just work."

R. G. Türeci et al., *Julia Programming for Physics Applications*,
https://doi.org/10.1007/978-3-031-84716-5_4

- **Concise**. Less code means fewer mistakes and more efficient development and analysis.
- **Flexible**. Produce your favourite plots from your favourite package, only quicker and simpler.
- **Consistent**. Don't commit to one graphics package. Use the same code and access the strengths of all backends.
- **Lightweight**. Very few dependencies, since backends are loaded and initialised dynamically.
- **Smart**. It's not quite AGI, but Plots should figure out what you want to do and what you tell it.

Plots is a high-level plotting package. It provides powerful graphing capabilities that are usually desired for most of the high-performance computing requirements. It interfaces with other plotting packages (referred to as *back ends* or *graphic engines*) to produce graphics files in a flexible manner. Each of these graphic back ends also can perform as stand-alone plotting packages, but Plots provides a user-friendly, simple, and consistent interface.

Before usage, a package must be imported into the present Julia session using the import *<packageName>* command.

A graphic object has many properties as below:

- **Size**: Length and breadth of a plot as it appears on a graphing terminal
- **Shape**: Aspect ratio
- **Title**: The string declaring the subject of a plot
- **Axes**: The axes that show reference data points
- **Labels**: The labels associated to axes showing a descriptive string
- **Markers**: Symbols of various shapes depicting data points
- **Resolution**: Measured in dpi (dots per inch) because, while printing, the resolution of an image is determined by the ability of the printer to define individual pixels as a dot and the number of dots per inch will define the degree of pixelation of the figure
- **Format of file**: A plot can be stored in many formats reserved for photographs and other kind of media
- **File name**: A file name provides identification marker for an object in a computing system.

Furthermore, a plot object is rendered by a graphical engine. Again, a variety of graphing engines of varying capabilities can be used. Some are more powerful in terms of their abilities to produce plots of richer properties and producing them faster. While working with different packages, some graphing engines provide a way to define a particular engine for rendering at the back end. Fixing this aspect alleviates worries about uniformity of configurations, quality, and formats of graphical objects.

The following is a brief list of some of the more commonly used Plots package functions for generating plots:

plot(): Can be used to plot data in various ways, including series data, single functions, and multiple functions, as well as for presenting and merging other plots. This is the most common plotting function.
scatter(): Used for plotting scattered data points not connected by a line.
bar(): Used for plotting bar graphs.
heatmap(): Used to plot a matrix, or an image.
surface(): Used to plot surfaces (3D plots). This is the typical way in which one would plot a real-valued function of two variables.
contour(): Used to create a contour plot. This is an alternative way to plot a real-valued function of two variables.
contourf(): Similar to contour(), but with shading between contour lines.
histogram(): Used to plot histograms of data.
stephist(): A stepped histogram. This is a histogram with no filling.

In addition, each of these functions also has a companion function with a "!" suffix, e.g. plot!(). These functions modify the previous plot, adding additional plotting aspects to them. This is shown in many examples throughout the book. Furthermore, the Plots package supplies additional important functions such as *savefig()* for saving a plot, annotate!() for adding annotations to plots, default() for setting plotting default arguments, and many more. Note that in the examples throughout this book *pyplot()* is called. This activates the PyPlot backend for plotting.

4.2 Line Plots

After you have installed Plots.jl via Pkg.add("Plots"), the first step is to initialise the package. This will take a few seconds depending on your computer properties. We can create a range from 0 to 20 of, say, 200 elements for the x coordinates, For the y coordinates, we can create a vector by evaluating cos(x) in an element-wise fashion. To do this in Julia, we insert a dot right after the function call in below code [2]. Finally, we use *plot()* to plot the line in Programme 4.1 (Fig. 4.1).

In Plots.jl, every column is a series, a set of related points which form lines, surfaces, or other plotting primitives. In Programme 4.2, we can plot multiple lines by plotting a matrix of values where each column is interpreted as a separate line. Below, [y1 y2] forms a 200 × 2 matrix (200 elements, 2 columns) [2] (Fig. 4.2).

```
[8]: using Plots
     x = range(0, 20, length=200)
     y = cos.(x)
     plot(x, y)
```

Programme 4.1 The code of cosine(x) plot

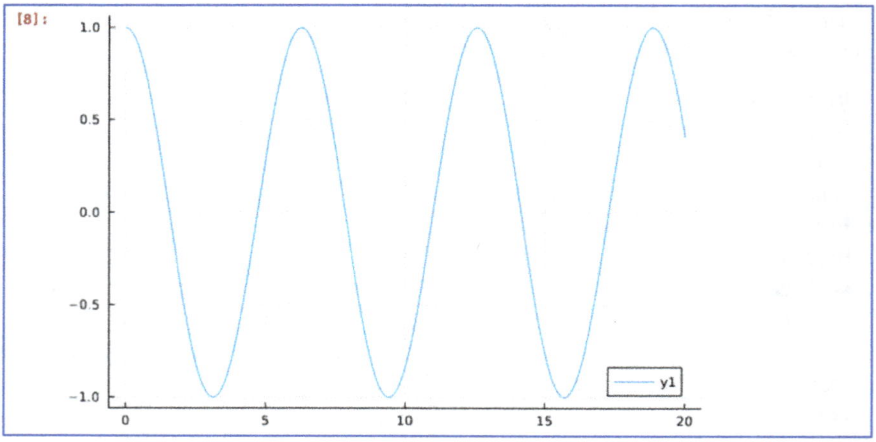

Fig. 4.1 The plot of the cosine(x)

```
[14]: using Plots
      x = range(0, 20, length=200)
      y1 = cos.(x)
      y2 = sin.(x)
      plot(x,[y1 y2])
```

Programme 4.2 The code of cosine(x) and sine(x) plot

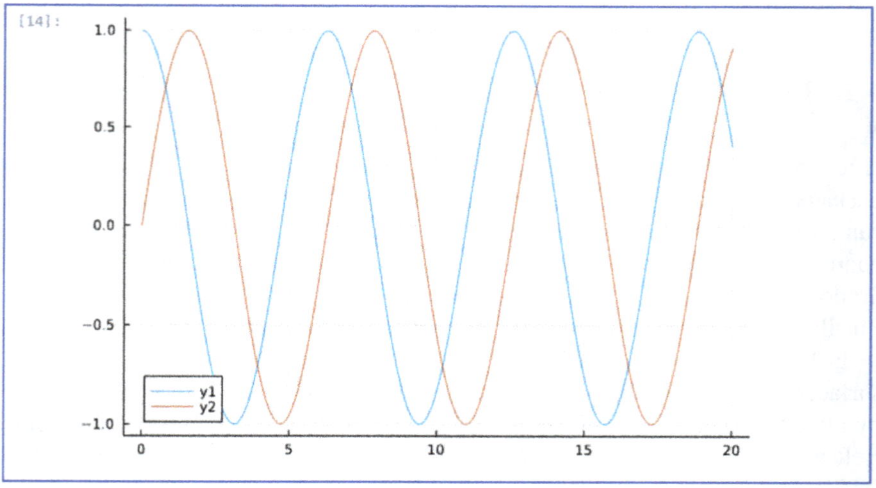

Fig. 4.2 The plots of the cos(x) and sin(x)

```
[20]: using Plots
      plot(x, [y1 y2])
      y3 = @. cos(x)^2 - 1/2    # equivalent to y3 = cos.(x).^2 .- 1/2
      plot!(x, y3)
```

Programme 4.3 The code of three plots using @. macro

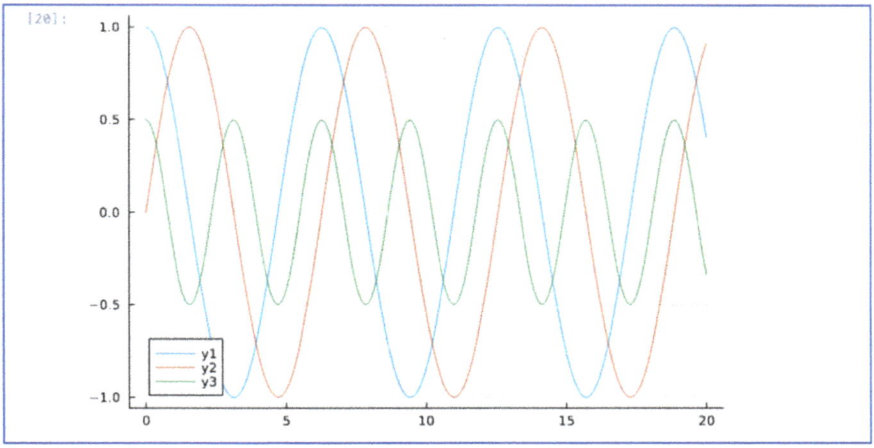

Fig. 4.3 Three plots using @. macro

Additionally, we can add more lines by mutating the plot object. This is done by the plot! command, where the ! denotes that the command is modifying the current plot. You will notice that we also use an @. macro. This is a convenience macro that inserts dots for every function call to the right of the macro, ensuring that the entire expression is to be evaluated in an element-wise manner. If we input the dots manually, we would need three of them for the sine, exponent, and subtraction, and the resulting code would be less readable. There is a code including three function [2] in Programme 4.3 (Fig. 4.3).

4.3 Plot Attributes

In Plots.jl, the modifiers to plots are called attributes [2]. Plots.jl follows two simple rules with data and attributes:

- Positional arguments correspond to input data
- Keyword arguments correspond to attributes.

We can change the line width using ***linewidth*** (or its alias lw), change the legend's labels using **label**, and add a title with ***title.*** Notice how ["cos(x)" "sin(x)"] has

```
[8]: using Plots
     x = range(0, 20, length=200)
     y1 = cos.(x)
     y2 = sin.(x)
     plot(x, [y1 y2], title="functions", label=["cos(x)" "sin(x)"], linewidth=4)
```

Programme 4.4 The code of cosine(x) and sine(x) plot with label and function

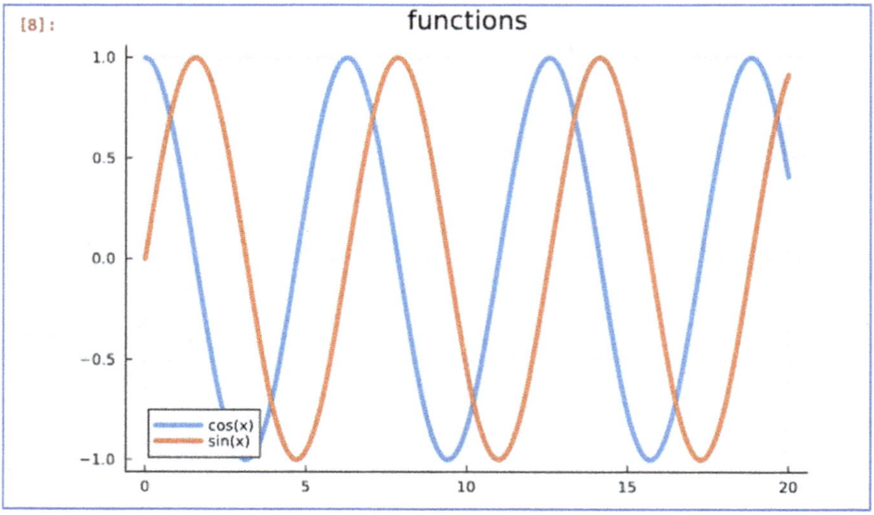

Fig. 4.4 The trigonometric functions plots with title and label

the same number of columns as the data. Additionally, since the line width is being attributed to [y1 y2], both lines will be affected by the *assigned* value in Programme 4.4 (Fig. 4.4).

Many attributes have their own dedicated modifier functions, while others can be accessed through *plot!(attribute = value)*. For instance, the xlabel attribute adds a label for the x-axis. We can specify it in the plot command with xlabel=... or we can use the modifier function below to add it after the plot has already been generated. You can decide which one is the best for code readability.

```
xlabel!("x")
```

Every modifier function is the name of the attribute followed by*!*. This will implicitly use the global Plots.current_plot. We can apply it to other plot objects via *attribute!(p, value)*, where p is the name of the plot object that wants to be modified.

The keywords and modifier functions interchangeably to perform some common modifications to our example, listed below. We will notice that for the attributes ls and legend, the values include a *colon*:. The colon denotes a symbol in Julia. They are commonly used for values of attributes in Plots.jl, along with strings and numbers.

- Labels for the individual lines, seen in the legend
- Line widths (we'll use the alias lw instead of linewidth)
- Line styles (we'll use the alias ls instead of *linestyle*)
- Legend position (outside the plot, as the default would clutter the plot)
- Legend columns (3 to better use the horizontal space)
- X-limits to go from 0 to 2pi
- Plot title and axis labels.

We can plot the three trigonometric functions plots Programme 4.5 with plot! in this code [2] (Fig. 4.5).

```
[16]: using Plots
      x = range(0, 20, length=200)
      y1 = cos.(x)
      y2 = sin.(x)
      y3 = @. cos(x)^2 - 1/2

      plot(x, [y1 y2], label=["cos(x)" "sin(x)"], lw=[2 1])
      plot!(x, y3, label="cos(x)^2 - 1/2", lw=4, ls=:dot)
      plot!(legend=:outerbottom, legendcolumns=3)
      xlims!(0, 2pi)
      title!("Functions")
      xlabel!("x")
      ylabel!("y")
```

Programme 4.5 The code of plot of three trigonometric functions

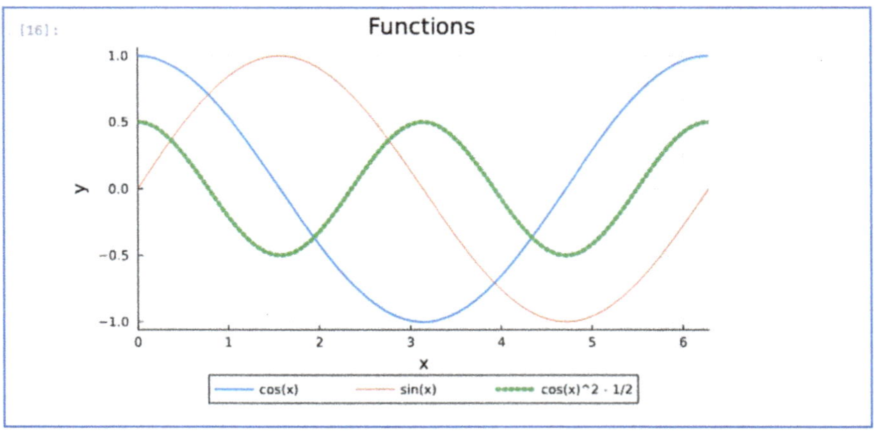

Fig. 4.5 The three trigonometric functions plots with plot!

4.4 Logarithmic Scale

A logarithmic scale is a method used to display numerical data which spans a broad range of values, when there are significant differences between the magnitudes of the numbers involved. If we need data needs to be plotted across orders of magnitude, we use logarithmic plots.

In Programme 4.6, the attributes *xscale* and yscale can be set to :log*10* in this case [2]. They can also be set to :identity to keep them linear-scale. The data and limits should be positive (Fig. 4.6).

```
[26]: using Plots
x = 10 .^ range(0, 5, length=200)
y = @. 1/(1+x)

plot(x, y, label="1/(1-x)")
plot!(xscale=:log10, yscale=:log10,minorgrid=true)
xlims!(1e+0, 1e+6)
ylims!(1e-5, 1e+0)
title!("Logplot")
xlabel!("x")
ylabel!("y")
```

Programme 4.6 The code of logarithmic plot

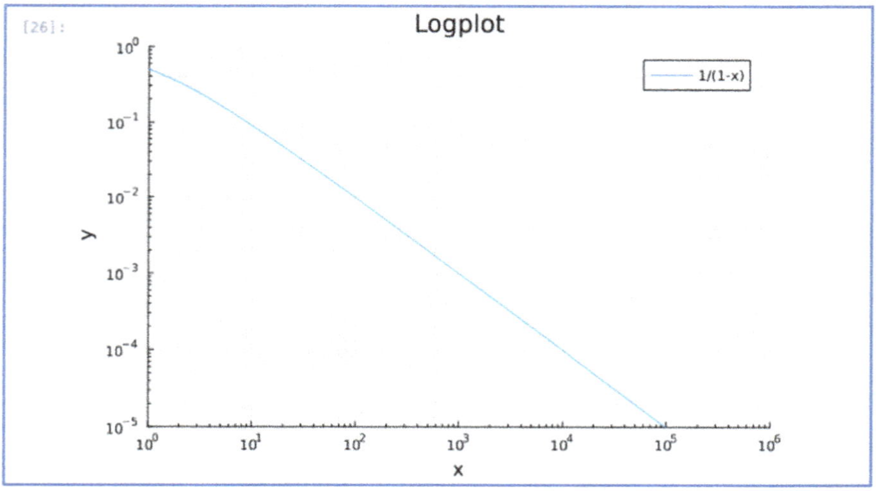

Fig. 4.6 The logarithmic plot

4.5 Histograms

Histograms are good for showing general distributional features of dataset variables. They are graphs where data are sampled into bins and numbers of data points belonging to particular bins are plotted. We can see roughly where the peaks of the distribution are, whether the distribution is skewed or symmetric, and if there are any outliers.

In Programme 4.7, plots functions have a method named ***histogram()*** that can be used as follows [3]. First, 1000 normalised random numbers are generated with the ***randn()*** function and then they are plotted using the **histogram**() function within Plots. The bell-shaped curve verifies that numbers are indeed normally distributed (Fig. 4.7).

4.5.1 Bar Charts

A bar chart is a graphical representation used to display and compare discrete categories of data through rectangular bars, where the length or height of each bar is proportional to the frequency or value of the corresponding category. In Programme 4.8, bar charts show vertical bars as per data for the y-axis using 20 random numbers [3] (Fig. 4.8).

```
[7]: using Plots
     x = randn(1000);
     Plots.histogram(x)
```

Programme 4.7 The code of histogram of random numbers generated using rand() function

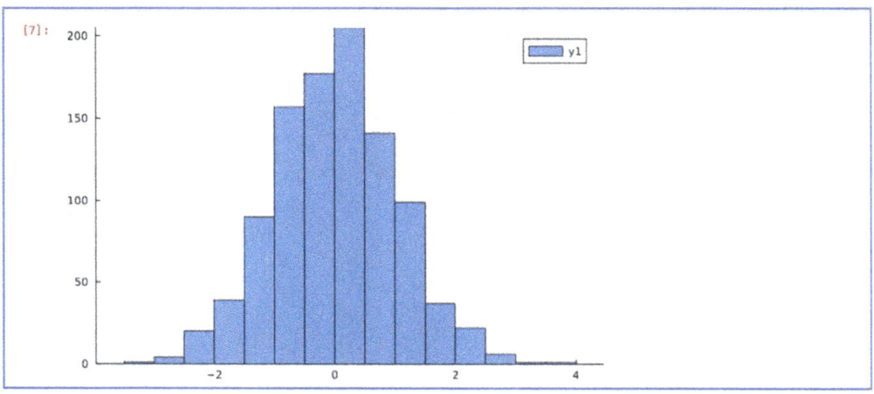

Fig. 4.7 Histogram of random numbers generated using rand() function

```
[18]: using Plots
      x = rand(50);
      Plots.bar(x)
```

Programme 4.8 The code of bar chart of 50 random numbers generated with rand() function

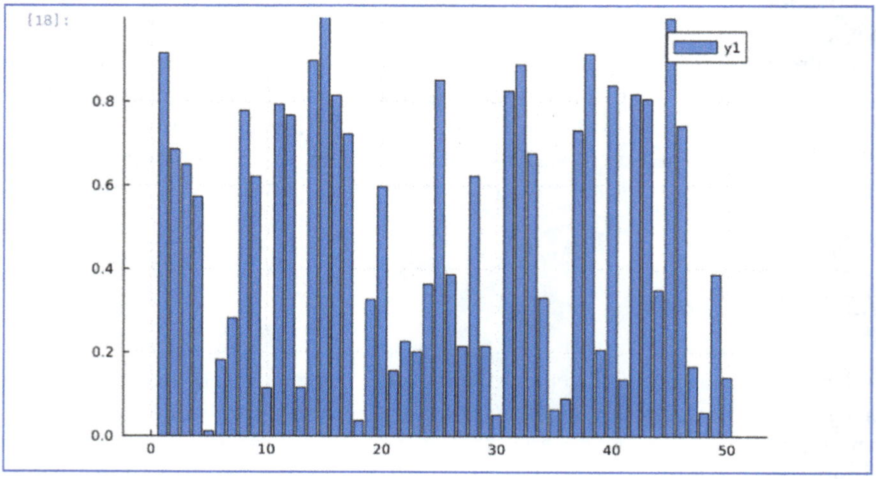

Fig. 4.8 Bar chart of 50 random numbers generated with rand() function

```
[42]: using Plots
      x = rand(20);
      Plots.pie(x)
```

Programme 4.9 The code of Pie Chart of 20 random numbers generated using rand() function

4.5.2 Pie Charts

A pie chart is a type of graph representing data in a circular form, with each slice of the circle representing a fraction or proportionate part of the whole. All slices of the pie add up to make the whole equaling 100% and 360° [4]. We can do an experiment with 20 random numbers using Programme 4.9 [3] (Fig. 4.9).

4.5.3 Scatter Plots

Scatter plots are the graphs that present the relationship between two variables in a dataset. It represents data points on a two-dimensional plane or on a Cartesian system. While the dependent variable is plotted on the y-axis, the independent variable or

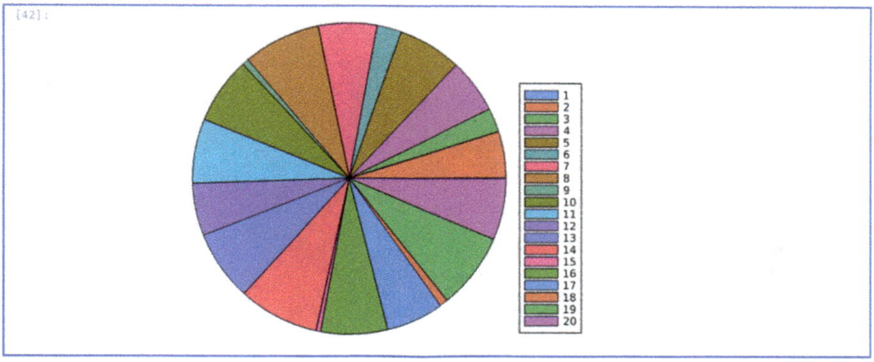

Fig. 4.9 Pie Chart of 20 random numbers generated using rand() function

```
[22]: using Plots
      x=1:200;
      y=randn(2000);
      Plots.scatter(x,y)
```

Programme 4.10 The code of scatter plot of 100 random numbers generated using the *rand()*

attribute is plotted on the x-axis. Scatter plots just put a dot for x and y data at the coordinate made by x and y.

In Programme 4.10, x-axis has linearly distributed 2000 numbers from 1 to 200 (y-axis has 200) random numbers, respectively, Nagar [3] (Fig. 4.10).

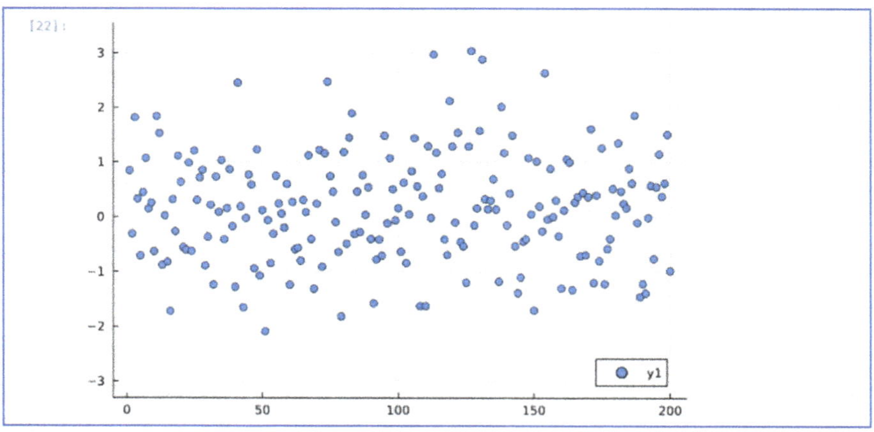

Fig. 4.10 Scatter plot of 100 random numbers generated using the *rand()*

```
[38]: using Plots
      x = randn(10^5)
      y = randn(10^5)
      histogram2d(x, y)
```

Programme 4.11 The code of 2D histogram

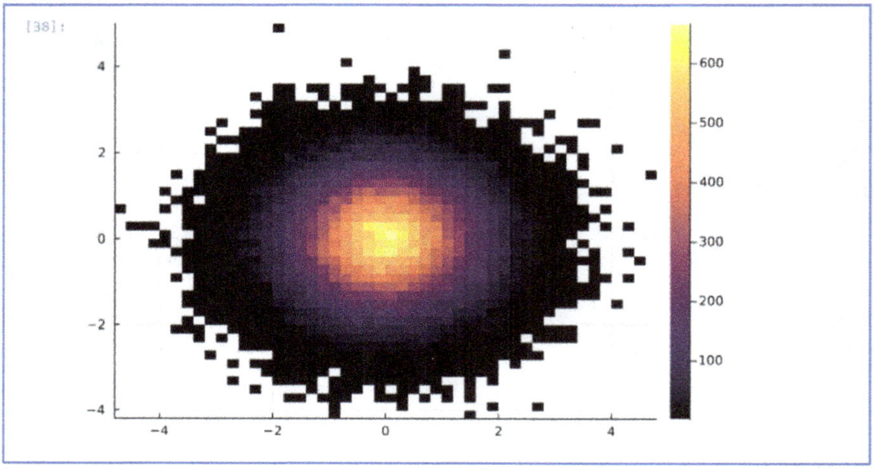

Fig. 4.11 2D histogram

4.5.4 2D Histograms

Two-dimensional histograms are accessed through the function histogram2d and its mutating variant histogram2d!. To plot them, two vectors x and y of the same length are needed [5].

Two-dimensional histograms are accessed through the function histogram2d and its mutating variant histogram2d!. To plot them, two vectors x and y of the same length are needed in Programme 4.11 (Fig. 4.11).

4.6 Normalization

Normalization can have a range of meanings in the statistics. It may refer to more sophisticated adjustments where the intention is to bring the entire probability distributions of adjusted values into alignment.

It is often desirable to normalise the histogram in some way. To do this, the normalise attribute is used, and we want *normalize = :pdf (or:true)* to normalise the total area of the bins to 1 in Programme 4.12. Since we sampled from the normal

```
[69]: using Plots
      p(x) = 1/sqrt(2pi) * exp(-x^2/2)
      b_range = range(-4, 4, length=27)

      histogram(x, label="Exp.", bins=b_range, normalize=:pdf, color=:red)
      plot!(p, label="Analy.", lw=3, color=:blue)
      xlims!(-10, 10)
      ylims!(0, 0.4)
      title!("Distribution, 10000 samples")
      xlabel!("x")
      ylabel!("P(x)")
```

Programme 4.12 The code of normalization histogram

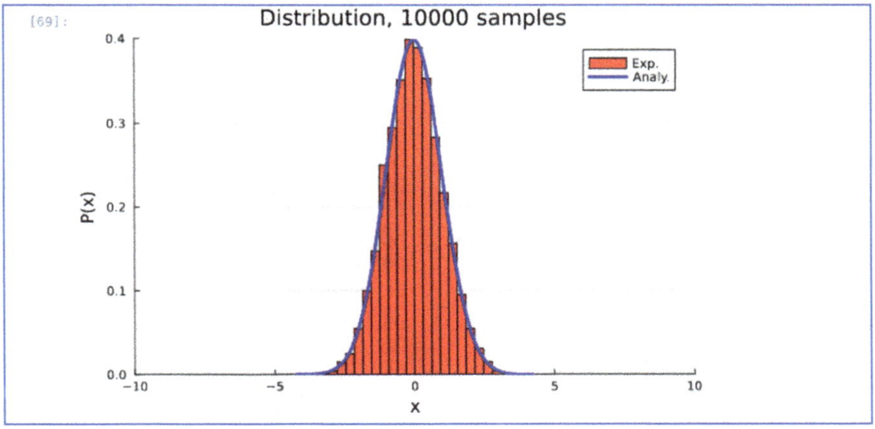

Fig. 4.12 The normalization histogram

distribution, we may as well plot it too. Of course, other common attributes like the title, axis labels, and colours can be changed as well [6] (Fig. 4.12).

4.7 Weighted Histograms

Another common feature is to weigh the values in x. We assume that x-consists of data sampled from a uniform distribution and we wanted to weigh the values according to an exponential function. We would pass in a vector of weights of the same length as x. To check that the weighting is done correctly, we plot the exponential function multiplied by a normalization factor in Programme 4.13 [6] (Fig. 4.13).

Note: ***Saving Figures***: For saving plots, we use the *savefig* command. There also exist convenience functions png, Plots.pdf and other unexported helpers. With these, the extension is omitted from the filename.

```
[83]: using Plots
      f_exp(x) = exp(x)/(exp(1)-1)

      x = rand(10^6)
      w = exp.(x)

      histogram(x, label="Exp.", bins=:scott, weights=w, normalize=:pdf, color=:red)
      plot!(f_exp, label="Analy.", lw=3, color=:black)
      plot!(legend=:topleft)
      xlims!(0, 1.0)
      ylims!(0, 1.6)
      title!("Uniform dist., weighted by exp(x)")
      xlabel!("x")
      ylabel!("P(x)")
```

Programme 4.13 The code of weighted histogram

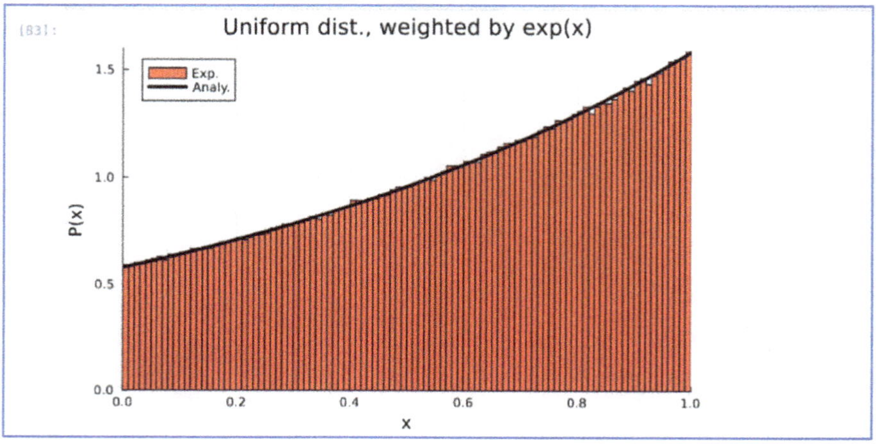

Fig. 4.13 The weighted histogram

References

1. https://docs.juliaplots.org/stable/
2. https://docs.juliaplots.org/stable/tutorial/#tutorial
3. Sandeep Nagar (2017), Beginning Julia Programming: For Engineers and Scientists
4. https://www.jaspersoft.com/articles/what-is-a-pie-chart
5. https://docs.juliaplots.org/latest/series_types/histogram/
6. https://docs.juliaplots.org/stable/series_types/histogram/#Normalization

Chapter 5
Symbolic Calculations with Julia

5.1 Introduction

Symbolic computing is possible in Julia as well as in programming languages such as Matlab, Wolfram Mathematica and Python. Thus, it is possible to perform analytic calculations on symbolically generated expressions; moreover, to solve, for example, an equation. One of the modules that can be used for symbolic computation is the *Symbolics* module [1]. This module contains some very useful features, which will be discussed in the following sections. Another useful module is the *SymEngine* module [2]. Another module is the *SymPy* module [3], which is designed for Python. This module can be also used as in Python with *PyCall* module [4]. However, it is also possible to perform extensive symbolic calculations with SymPy without the need for the *PyCall* module.

In this section, we will first discuss the symbolic calculations that can be done with the Symbolics module. In the next step, we will discuss calculations with the *SymEngine* module. Our main interest will be on calculations with the *SymPy* module.

5.2 Symbolic Calculations with Symbolics Module

In object-oriented programming, every quantity is an object. These objects must also be defined as a data type. In this respect, symbolic calculation is a calculation performed with objects defined symbolically without assigning numeric values. In this section we will discuss the calculations that can be done with the *Symbolics* module. Symbolic objects or variables are defined in this module with the ***@variables*** macro as in Programme 5.1.

After defining a variable, Programme 5.2, a function can be defined depending on this variable. With the ***substitute()*** function, a value can be assigned to the variable and thus the numerical value of the function can be determined. In the seventh cell of

R. G. Türeci et al., *Julia Programming for Physics Applications*,
https://doi.org/10.1007/978-3-031-84716-5_5

```
[1]: using Symbolics

[2]: @variables x

[2]:
```

$$[x]$$

```
[3]: @variables y z

[3]:
```

$$\begin{bmatrix} y \\ z \end{bmatrix}$$

Programme 5.1 Defining in one or more variables in Julia

the programme, the function ***f1*** is defined with the variable x. In the next two cells, the value of the function is calculated by assigning a value to x with the ***substitute()*** function. In the next cell, the ***map()*** function was used to define a list with the values of the function ***f1***, but the desired result was not obtained. The correct usage is given in the 15th cell.

There is an important detail here. Programme 5.2, function ***f1*** is a function containing only one variable. If the function contains more than 1 variable, then the ***substitute()*** function must be defined in Dictionary data type. An example is given in Programme 5.3.

There is an important detail here. Programme 5.2, function ***f1*** is a function containing only one variable. If the function contains more than 1 variable, then the ***substitute()*** function must be defined in Dictionary data type. An example is given in Programmes 5.3 and 5.4.

The Symbolics module contains the functions ***simplify()*** to simplify symbolic expressions and ***expand()*** to obtain the most explicit form of the expression. Some applications on trigonometric expressions can be seen in Programme 5.5.

5.2.1 Derivative with Symbolics Module

Expressions defined with symbolic variables can be differentiated with the ***Symbolics.derivative*** function. If this derivative is assigned to a variable, its numerical value can also be calculated using the *substitute()* function mentioned above. Moreover, higher order derivatives can be calculated by applying the derivative function one after the other (Programme 5.6).

Let's summarise Programme 5.7. In cell 23, variables and function ***f8*** are defined with these variables. In the last line of the cell, the function ***Symbolics.derivative()*** is used to take the derivative of function ***f8*** with respect to variable x. In the next cell, the derivative of the f8 function with respect to the y variable is taken. In the 25th cell of the programme, the results are printed with ***println()***. In the next cell, the value of ***x*** in the derivative expression is assigned as 2 and in the 27th cell, both

```
[4]: f1 = x^2
```

[4]: x^2

```
[5]: value_f1 = substitute(f1, (x => 3))
```

[5]: 9

```
[6]: value2_f1 = substitute(f1, (x => sqrt(3)))
```

[6]: 3

```
[7]: x_values = 0:1:4
     f_values = map(x -> f1, x_values)
```

[7]:

$$\begin{bmatrix} x^2 \\ x^2 \\ x^2 \\ x^2 \\ x^2 \end{bmatrix}$$

```
[8]: f_values = map(i -> substitute(f1, (x => i)), x_values)
```

[8]:

$$\begin{bmatrix} 0 \\ 1 \\ 4 \\ 9 \\ 16 \end{bmatrix}$$

Programme 5.2 Assigning a value to a variable

```
[9]: @variables x y z

     f2 = 3x^2 + 4log(y) - exp(-z)
```

[9]: $4\log(y) - e^{-z} + 3x^2$

```
[10]: f2_value = substitute(f2, Dict(x => 2, y => 2, z => 3))
```

[10]: 14.723

Programme 5.3 Assigning a value set to the variables by Dictionary data type

x and ***y*** values are given as ***2*** and the numerical value of the derivative expression is calculated for these values. Programme 5.8, higher order derivatives are taken.

The name of this function is relatively long. Fortunately, in Programme 5.9, we can perform higher order derivatives in a shorter form by defining ***df* = *Symbolics.derivative***.

```
[11]: f3 = sin(x)^2 + cos(y)^2

      f3_value = substitute(f3, Dict(x => deg2rad(30), y => deg2rad(30)))
```

[11]: 1

```
[12]: f4 = sin(x)^2 + cos(y)^2

      f4_value = substitute(f3, Dict(x => pi/6, y => pi/6))
```

[12]: 1

```
[13]: f5 = f2 + f4

      f2_value = substitute(f2, Dict(x => pi/4, y => pi/3, z => 1))
```

[13]: 1.6671

```
[14]: f5
```

[14]: $4\log\left(y\right) - e^{-z} + 3x^2 + \cos^2\left(y\right) + \sin^2\left(x\right)$

Programme 5.4 Assigning a value set to the variables by Dictionary data type

```
[15]: f6 = sin(x)^2 + cos(x)^2
```

[15]: $\sin^2\left(x\right) + \cos^2\left(x\right)$

```
[16]: simplify(f6)
```

[16]: 1

```
[17]: f7 = 2sin(x)cos(x)
```

[17]: $2\sin\left(x\right)\cos\left(x\right)$

```
[18]: simplify(f7)
```

[18]: $\sin\left(2x\right)$

```
[19]: expand(f7)
```

[19]: $2\sin\left(x\right)\cos\left(x\right)$

Programme 5.5 *Simplify()* and *expand()* functions in symbolics

```
[20]: f8 = (x+1)^3

      simplify(f8)
```

[20]:

$$(1 + x)^3$$

```
[21]: expand(f8)
```

[21]:

$$1 + 3x + 3x^2 + x^3$$

Programme 5.6 *Simplify()* and *expand()* functions in symbolics

```
[23]: @variables x y
      f8 = x^2 + y^3 - sin(x * y)
      df_x = Symbolics.derivative(f8, x)
```

[23]:

$$2x - y\cos(xy)$$

```
[24]: df_y = Symbolics.derivative(f8, y)
```

[24]:

$$-x\cos(xy) + 3y^2$$

```
[25]: println("Expression : \t\t\t $f8")
      println("Derivative with respect to x :\t $df_x")
      println("Derivative with respect to y :\t $df_y")

      Expression :                      -sin(x*y) + x^2 + y^3
      Derivative with respect to x :    2x - y*cos(x*y)
      Derivative with respect to y :    -x*cos(x*y) + 3(y^2)
```

```
[26]: df_x_value = substitute(df_x, x => 2)
```

[26]:

$$4 - y\cos(2y)$$

```
[27]: df_x_value = substitute(df_x, Dict(x => 2, y => 2))
```

[27]:

5.3073

Programme 5.7 The derivation of a function with Symbolics

We can draw the graphs of the functions and derivatives we work on. But for this we need to do an intermediate operation. When we calculate the values of the function and its derivatives in the range defined as ***x_values***, these values are kept symbolically. Therefore, we need to convert these values to numeric values in a way that Julia can understand. ***Symbolic.value()*** function is used for this purpose. Programme 5.10, first the numeric values of the function and its derivatives are calculated with ***substitute()***, and in the last five lines these calculations are converted from symbolic to numeric values. The ***Plots*** module can now be called to plot the function and its derivatives, Programme 5.11. Programme 5.10, the semicolon (***;***)

```
[28]: df_x2 = Symbolics.derivative(df_x, x)
```

[28]:

$$2 + y^2 \sin(xy)$$

```
[29]: df_x3 = Symbolics.derivative(df_x2, x)
```

[29]:

$$y^3 \cos(xy)$$

```
[30]: df_y2 = Symbolics.derivative(df_y, y)
```

[30]:

$$6y + x^2 \sin(xy)$$

```
[31]: df_y3 = Symbolics.derivative(df_y2, y)
```

[31]:

$$6 + x^3 \cos(xy)$$

Programme 5.8 High order derivatives

```
[33]: df(f8, x)
```

[33]:

$$2x - y\cos(xy)$$

```
[34]: df(df(f8,x), x)
```

[34]:

$$2 + y^2 \sin(xy)$$

```
[35]: df(df(df(f8,x), x), x)
```

[35]:

$$y^3 \cos(xy)$$

```
[36]: df(f8, y)
```

[36]:

$$-x\cos(xy) + 3y^2$$

Programme 5.9 Higher order derivatives using a long function name with its short name

symbol used in the last line is used to prevent the results from being printed on the screen (Fig. 5.11a).

Another function to be used for derivative operation is ***Differential()*** function. This function takes the variable as a parameter according to which derivative operation will be performed. Programme 5.12, the derivative operator depending on ***x*** is defined by ***Dx*** and the derivative of the function ***f8*** is taken. However, this derivative is given in a symbolic notation. In order to see the explicit form of the derivative, the ***expand_derivatie()*** function should be used.

```
[39]: x_values = -4:0.1:4

      f8_values = [substitute(f8, Dict(x => i, y => i)) for i in x_values]
      dfx1_val = [substitute(dfx1, Dict(x => i, y => i)) for i in x_values]
      dfx2_val = [substitute(dfx2, Dict(x => i, y => i)) for i in x_values]
      dfy1_val = [substitute(dfy1, Dict(x => i, y => i)) for i in x_values]
      dfy2_val = [substitute(dfy2, Dict(x => i, y => i)) for i in x_values]

      f8_numeric = [Symbolics.value(i) for i in f8_values]
      dfx1_num = [Symbolics.value(i) for i in dfx1_val]
      dfx2_num = [Symbolics.value(i) for i in dfx2_val]
      dfy1_num = [Symbolics.value(i) for i in dfy1_val]
      dfy2_num = [Symbolics.value(i) for i in dfy2_val];
```

Programme 5.10 Convert the symbolic values to the numerical values

```
[40]: using Plots

      p = plot(x_values, f8_numeric, label="f8(x,y) = $f8", legend=:top)
      plot!(x_values, dfx1_num, label="dfx1(x,y) = $dfx1")
      plot!(x_values, dfx2_num, label="dfx2(x,y) = $dfx2")
      plot!(x_values, dfy1_num, label="dfy1(x,y) = $dfy1")
      plot!(x_values, dfy1_num, label="dfy2(x,y) = $dfy2")

      xlabel!("x"); ylabel!("y"); title!("Graphs of f8 and its derivations")
```

Programme 5.11 The graphs of the function and its derivations (the graph is on the next page)

[40]:

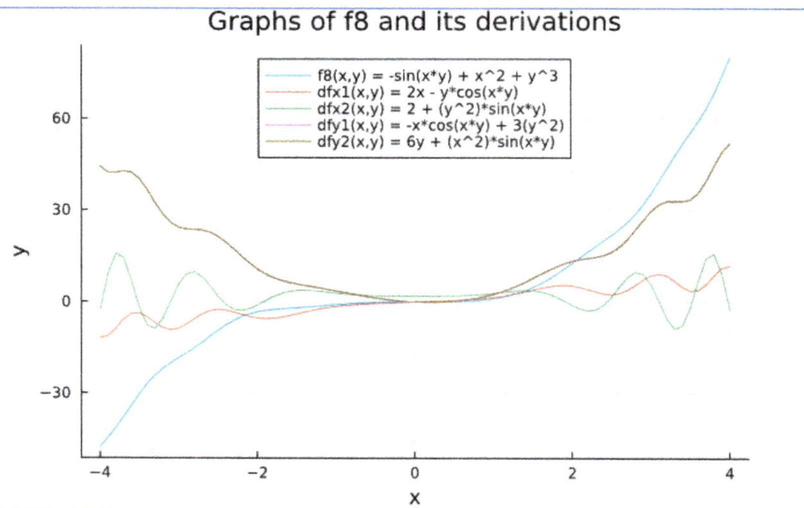

Programme 5.11a The graphs of the function and its derivations

```
[42]: Dx = Differential(x)
      Dy = Differential(y)
      f8
```

[42]:

$$-\sin(xy) + x^2 + y^3$$

Programme 5.12 *Differential()* function

These operations can be considered as partial derivatives. Therefore, in order to calculate the exact derivative or gradient of a function, these partial derivatives can be added together, and these derivatives can be saved in a list and defined in vector form. For higher order derivatives, the variable parameter of the ***Differential()*** function should be taken, and this should be specified as the number of exponents of which order to take the derivative.

Programme 5.12 shows the definition of the ***Differential()*** function for the variables ***x*** and ***y*** and the function f8 is called. Programme 5.13 shows the calculation of these derivatives. However, as can be seen in cells 44 and 47, the function ***expand_derivatives()*** must be used to print the explicit derivatives.

```
[43]: dfx1_D = Dx(f8)
```

[43]:

$$\frac{\mathrm{d}}{\mathrm{d}x}\left(-\sin(xy) + x^2 + y^3\right)$$

```
[44]: expand_derivatives(dfx1_D)
```

[44]:

$$2x - y\cos(xy)$$

```
[45]: dfy1_D = Dy(f8)
```

[45]:

$$\frac{\mathrm{d}}{\mathrm{d}y}\left(-\sin(xy) + x^2 + y^3\right)$$

```
[46]: expand_derivatives(dfy1_D)
```

[46]:

$$-x\cos(xy) + 3y^2$$

```
[47]: dfx1_D + dfy1_D
```

[47]:

$$\frac{\mathrm{d}}{\mathrm{d}y}\left(-\sin(xy) + x^2 + y^3\right) + \frac{\mathrm{d}}{\mathrm{d}x}\left(-\sin(xy) + x^2 + y^3\right)$$

```
[48]: expand_derivatives(dfx1_D + dfy1_D)
```

[48]:

$$2x - x\cos(xy) + 3y^2 - y\cos(xy)$$

Programme 5.13 *Differential()* function and its application

To calculate higher order derivatives with the ***Differential()*** function, the order of the function is taken as ***Differential()^n*** (where n is an integer). In Programme 5.14, an application of this is made and the third derivative of the function ***f8*** is calculated.

Note that the Greek alphabet can be used in Julia. For example, by typing ***\alpha*** or ***\beta*** and then pressing the ***TAB*** key, a small window will appear with the appropriate Greek alphabet, which can be selected to include the desired character in the programme, Programmes 5.14 and 5.15.

t and γ are defined as variables. $\in$ (..) defines a function with an unspecified argument, which can be used later with a specific argument in Programme 5.16. $\beta[1:3]$ will write the parameter beta with subindex from 1 to 3. On the next line, $w(..)$ is again a function to be used later with an argument, $x(t)$ specifies that x is a function of t and $z(t, x)$ specifies that z is a function of t and $x(t)$.

Programme 5.17 defines a vector with three elements ***u*** [1], ***u*** [2] and ***u*** [3]. The elements of this vector are a symbolic array with ***Symbolics.Arr{Num, 1}*** indicating that it consists of elements of type Num of dimension 1. The function ***Symbolics.scalarize()*** makes symbolic expressions more understandable and provides flexibility for symbolic arrays and matrices.

```
[ ]: \beta
     β
```

Programme 5.14 Greek alphabet characters in Julia

```
[1]: β

UndefVarError: `β` not defined

[2]: using Symbolics

     @variables β

[2]: [β]
```

Programme 5.15 Greek alphabet characters in Julia

```
[4]: @variables t γ ϵ(..) β[1:3]
     @variables w(..) x(t) y z(t, x)

     equation = β[3]*tanh(x) - β[1]* x + y^γ + γ*ϵ(3) + (z - t) - β[2] * w(t - γ)
```

[4]:

$$-t + z\left(t, x\left(t\right)\right) + y^{\gamma} + \tanh\left(x\left(t\right)\right)\beta_3 - x\left(t\right)\beta_1 - w\left(t - \gamma\right)\beta_2 + \epsilon\left(3\right)\gamma$$

```
[5]: log(equation)
```

[5]:

$$\log\left(-t + z\left(t, x\left(t\right)\right) + y^{\gamma} + \tanh\left(x\left(t\right)\right)\beta_3 - x\left(t\right)\beta_1 - w\left(t - \gamma\right)\beta_2 + \epsilon\left(3\right)\gamma\right)$$

Programme 5.16 An equation with different variables

```
[7]: @variables u[1:3] #defines a vector

[7]: 1-element Vector{Symbolics.Arr{Num, 1}}:
      u[1:3]

[8]: Symbolics.scalarize(u)

[8]:
```

$$\begin{bmatrix} u_1 \\ u_2 \\ u_3 \end{bmatrix}$$

```
[9]: @variables y[1:3, 1:6] # defines a tensor

[9]: 1-element Vector{Symbolics.Arr{Num, 2}}:
      y[1:3,1:6]

[10]: Symbolics.scalarize(y)

[10]:
```

$$\begin{bmatrix} y_{1.1} & y_{1.2} & y_{1.3} & y_{1.4} & y_{1.5} & y_{1.6} \\ y_{2.1} & y_{2.2} & y_{2.3} & y_{2.4} & y_{2.5} & y_{2.6} \\ y_{3.1} & y_{3.2} & y_{3.3} & y_{3.4} & y_{3.5} & y_{3.6} \end{bmatrix}$$

```
[11]: @variables r k(r)[1:3]

[11]: 2-element Vector{Any}:
       r
        (k(r))[1:3]

[12]: Symbolics.scalarize(k)

[12]:
```

$$\begin{bmatrix} k(r)_1 \\ k(r)_2 \\ k(r)_3 \end{bmatrix}$$

Programme 5.17 Different variable definitions in Julia with symbolics module

In Programme 5.18, different variable definition methods are given. In cell 13, ***runtime_symbol_value*** and ***:value_b*** and ***:value_c*** are defined. These are used to define variable names at runtime and are a good example of the flexibility of the Julia language. In the next cell ***t*** is a symbolic variable. The ***$*** sign works like string interpolation, that is, it places the value of the variable so that it takes value at runtime. ***$b(t)*** defines a function that depends on t, using the contents of the variable ***b***. ***$c(t)[1:3]*** is used to define a vector function with three indexed elements depending on ***t***.

```
[13]: a, b, c = :runtime_symbol_value, :value_b, :value_c

[13]: (:runtime_symbol_value, :value_b, :value_c)

[14]: vars = @variables t $a $b(t) $c(t)[1:3]

[14]: 4-element Vector{Any}:
           t
       runtime_symbol_value
         value_b(t)
             (value_c(t))[1:3]
```

Programme 5.18 Different variable definitions in Julia with symbolics module

5.2.1.1 Symbolic Linear Algebra with Symbolics Module

With the inclusion of the *LinearAlgebra* module [5] in the programme, linear algebra operations such as determinant of a matrix, inverse of a matrix, eigenvalues and eigenvectors of a matrix can be performed symbolically. In some cases, for example, to determine the eigenvalues and eigenvectors of a matrix, a value must be assigned to the symbolic matrix. Programme 5.19, a 3×3 matrix is defined and the determinant of this matrix is calculated symbolically. Then the numerical value of this determinant is determined by assigning values to variables.

In Programme 5.20, the matrix defined in the previous programme is defined with another name and the eigenvalues and eigenvectors of matrix B are calculated after assigning values to the variables. ***eigen()*** function is used for this purpose. Programme 5.10, the inverse of matrix B is computed and the ***inv()*** function is used for this purpose. As it is known, the product of a matrix and this matrix gives the unit matrix, Programme 5.22. However, matrix multiplication is also possible symbolically by using the *LinearAlgebra* and *Symbolics* modules together, Programme 5.23 (Programme 5.21).

However, dot and cross functions are used for scalar and vector product of two vectors respectively, Programme 5.24.

5.3 Symbolic Calculations with SymEngine Module

The ***SymEngine*** module is a C++ module for fast symbolic operations. Although this module is written in C++, it can also be used in R, Julia, Python, Ruby and Haskell. The module can be used to take symbolic derivatives and integrals. Mathematical

```
[24]: using Symbolics
      using LinearAlgebra

[25]: @variables x y z

      A = [x y 2z; 3x y z; x y z]
```

[25]:

$$\begin{bmatrix} x & y & 2z \\ 3x & y & z \\ x & y & z \end{bmatrix}$$

```
[26]: det_A = det(A)
```

[26]:

$$2xyz$$

```
[27]: det_A_val = substitute(det_A, Dict(x => 2, y => 2, z => 1))
```

[27]:

$$8$$

Programme 5.19 Symbolic matrix and the determinant of this matrix

```
[28]: B = [x y 2z; 3x y z; x y z]

      B_num = substitute(B, Dict(x => 1.0, y => 1.0, z => 1.0))
      eigenvals, eigenvecs = eigen(B_num)

[28]: Eigen{ComplexF64, ComplexF64, Matrix{ComplexF64}, Vector{ComplexF64}}
      values:
      3-element Vector{ComplexF64}:
       -0.4502859374559802 - 0.5567655710868958im
       -0.4502859374559802 + 0.5567655710868958im
        3.9005718749119636 + 0.0im
      vectors:
      3x3 Matrix{ComplexF64}:
       0.320126+0.244247im  0.320126-0.244247im   0.54667+0.0im
       -0.83789-0.0im       -0.83789+0.0im        0.71542+0.0im
       0.254802-0.266231im  0.254802+0.266231im  0.435117+0.0im
```

Programme 5.20 Eigenvalues and eigenvectors of B matrix

```
[29]: B
```

[29]:

$$\begin{bmatrix} x & y & 2z \\ 3x & y & z \\ x & y & z \end{bmatrix}$$

```
[30]: B_inverse = inv(B)
```

[30]:

$$\begin{bmatrix} 0 & \frac{1}{2x} & \frac{-1}{2x} \\ \frac{-1}{y} & \frac{-1}{2y} & \frac{5}{2y} \\ \frac{1}{z} & 0 & \frac{-1}{z} \end{bmatrix}$$

Programme 5.21 The inverse of B matrix

```
[71]: B*inv(B)
```

[71]:

$$\begin{bmatrix} 1 & 0 & 0 \\ 0 & 1 & 0 \\ 0 & 0 & 1 \end{bmatrix}$$

Programme 5.22 The unit matrix by multiplying a matrix and its inverse

```
[72]: B*B
```

[72]:

$$\begin{bmatrix} x^2 + 3xy + 2xz & xy + y^2 + 2yz & 2xz + yz + 2z^2 \\ 3x^2 + 3xy + xz & 3xy + y^2 + yz & 6xz + yz + z^2 \\ x^2 + 3xy + xz & xy + y^2 + yz & 2xz + yz + z^2 \end{bmatrix}$$

Programme 5.23 The matrix multiplication

```
[73]: v1 = [1, 2, 3]
      v2 = [4, 5, 6]

      result = dot(v1, v2) # scaler product

[73]: 32

[74]: v1 = [1, 2, 3]
      v2 = [4, 5, 6]

      result = cross(v1, v2) # cross product

[74]: 3-element Vector{Int64}:
       -3
        6
       -3
```

Programme 5.24 Scalar and vector multiplication of two vectors

expressions can be simplified, serialisation can be performed, moreover, algebraic differential equations can be solved.

Let's start with variable definition. It is possible to define variables with three different methods in *SymEngine*. In Programme 5.25 one variable and more than one variable are defined. In the first two methods, the ***symbols()*** function is used. In the symbols function, the variable can be defined either with a colon or as a string. The third method is to use the @***vars*** macro.

In this module, ***subs()*** function is used to assign numeric values or another variable to variables. But this operation does not change the function, only the number value

```
[2]: using SymEngine

[3]: x = symbols(:x)

[3]: x

[4]: x = symbols("x")

[4]: x

[5]: @vars x

[5]: (x,)

[6]: x, y = symbols(:x), symbols(:y)

[6]: (x, y)

[7]: x, y = symbols("x y")

[7]: (x, y)

[8]: @vars x y

[8]: (x, y)
```

Programme 5.25 Symbol definition methods in *SymEngine*

is written. For this reason, it is necessary to define another variable name to hold the expression that has been assigned a numeric value, Programme 5.26.

```
[10]: @vars α β
[10]: (α, β)
[11]: f1 = α^2 + β^2
[11]: α^2 + β^2
[12]: subs(f1, α, 4)
[12]: 16 + β^2
[13]: f1 # f1 doesn't change
[13]: α^2 + β^2
[14]: f1_α_val = subs(f1, α, 4)
[14]: 16 + β^2
[15]: f1_val = subs(f1, (α, 4), (β, 5))
[15]: 41
[16]: f1_val = subs(f1, Dict(α => 4, β => 5))
[16]: 41
[17]: f1(4, 5)
[17]: 41
[18]: f1(x, y)
[18]: x^2 + y^2
```

Programme 5.26 Different assigning methods in *SymEngine*

```
[19]: γ = symbols(:γ)
      f2 = subs(f1, (α, cos(γ)), (β, sin(γ)))
      f2
[19]: sin(γ)^2 + cos(γ)^2
[20]: f2(α)
[20]: sin(α)^2 + cos(α)^2
[21]: f2(pi/6)
[21]: 1.0
[22]: f2
[22]: sin(γ)^2 + cos(γ)^2
```

Programme 5.27 Assigning another variable to variable of a function

```
[22]: f2

[22]: sin(y)^2 + cos(y)^2

[23]: angles_in_deg = 0:30:180
      angles_in_rad = map(deg2rad, angles_in_deg)

      result = [subs(f2, y => i) for i in angles_in_rad]

      println(result)

      Basic[1.0, 1.0, 1.0, 1.0, 1.0, 1.0, 1.0]

[24]: result

[24]: 7-element Vector{Basic}:
       1.0
       1.0
       1.0
       1.0
       1.0
       1.0
       1.0

[25]: typeof(result)

[25]: Vector{Basic} (alias for Array{Basic, 1})

[26]: result2 = [subs(f2, y => i) |> N for i in angles_in_rad]

      println(result2)

      [1.0, 1.0, 1.0, 1.0, 0.9999999999999999, 1.0, 1.0]

[27]: typeof(result2)

[27]: Vector{Float64} (alias for Array{Float64, 1})
```

Programme 5.28 Assigning another variable to variable of a function

If the function depends on more than one variable and values or other variables are to be assigned to these variables, the values to be assigned to the variables can be written separately in parentheses or numeric values or different variable name can be given with the Dict data type as in the Symbolics module. However, the value of the function can also be calculated by opening parenthesis after the name of the function and writing the values to be assigned to the variables in this parenthesis (Programme 5.27).

In Programme 5.28, a number array named ***angles_in_deg*** is created. By applying ***deg2rad*** function in ***map()*** function to this array of numbers, angle values in degrees are converted to radians. In the next step, a list data named result is created with *list comprehension*. But the result is given as a Basic number array. This is a data type belonging to the *SymEngine* module and this array must be converted to numbers. When checked with the ***typeof()*** function, the result ***Vector{Basic}*** appears.

In line 26 of the programme, the pipe operator (|> operator) is used to perform the operation in cell 23 of the programme. This operator is used in Julia to apply one function to the result of another function. Here the result of the ***subs*** function is applied to the ***N*** function. Thus, the result is obtained as an array of numbers as expected, not as an array of Basic numbers.

SymEngine module allows to series expressions with the ***series*** function. It is known that when a function is expanded to a series around $\boldsymbol{x = 0}$, this is a Taylor series. This function takes 4 parameters. The first parameter is the name of the function to be expanded to the series. The second parameter is the name of the variable. The third parameter is a numerical value indicating where the series will be expanded. If this value is chosen as zero, the series is a Taylor series. The last parameter is again a numeric value related to how many elements of the series will be printed. In Programme 5.29, first the variable x is defined symbolically, then the function ***sin(x)*** is defined with this variable. The programme opens the Taylor series of this function.

Taylor expansion of ***cos(x)*** is given in Programme 5.30. In the same programme, Programme 5.29 shows various operations performed with the Taylor expansion of the ***sin(x)*** function. In Programme 5.31, the Taylor expansion of the exponential function is given.

In Programme 5.32, the Taylor expansion of the function ***sin(x)*** for $n = 3$, 5 and 7 was calculated. In Programme 5.33, the graph of the function ***sin(x)*** is plotted on a defined interval. In the next programme, Programmes 5.32 and 5.33 are combined and the graph of the ***sin(x)*** function and its Taylor expansion are plotted.

```
[29]: x = symbols(:x)
      f = sin(x)
      f_Taylor_exp = series(f, x, 0, 19)

[29]: x + (-1/6)*x^3 + (1/120)*x^5 + (-1/5040)*x^7 + (1/362880)*x^9 + (-1/39916800)*x^11 + (1/6227020800)
      *x^13 + (-1/1307674368000)*x^15 + (1/355687428096000)*x^17 + (-1/121645100408832000)*x^19
```

Programme 5.29 Taylor series expansion of sin(x)

```
[30]: g = cos(x)

      g_Taylor_exp = series(g, x, 0, 18)

[30]: 1 + (-1/2)*x^2 + (1/24)*x^4 + (-1/720)*x^6 + (1/40320)*x^8 + (-1/3628800)*x^10 + (1/479001600)*x^12
      + (-1/87178291200)*x^14 + (1/20922789888000)*x^16 + (-1/6402373705728000)*x^18

[31]: h = f_Taylor_exp + g_Taylor_exp

      h(deg2rad(45))

[31]: 1.41421356237309

[32]: h(pi/4)

[32]: 1.41421356237309

[33]: f1 = f_Taylor_exp^2 + g_Taylor_exp^2

      f1(deg2rad(pi/4))

[33]: 1.0
```

Programme 5.30 The Taylor expansion of cos(x) function

```
[34]: w = exp(x)

      w_Taylor_exp = series(w, x, 0, 7)

[34]: 1 + x + (1/2)*x^2 + (1/6)*x^3 + (1/24)*x^4 + (1/120)*x^5 + (1/720)*x^6 + (1/5040)*x^7

[35]: w_Taylor_exp(1)

[35]: 685/252

[36]: float(w_Taylor_exp(1))

[36]: 2.7182539682539684

[37]: exp(1)

[37]: 2.718281828459045
```

Programme 5.31 The Taylor expansion of exp(x) function

```
[38]: f = sin(x)
      n = [3, 5, 7]
      taylor_f = Dict()

      for i in n
          taylor_f[i] = series(f, x, 0, i)
      end

      for (i, series_expr) in taylor_f
          println("Taylor Series (n: $i): ", series_expr)
      end

      Taylor Series (n: 5): x + (-1/6)*x^3 + (1/120)*x^5
      Taylor Series (n: 7): x + (-1/6)*x^3 + (1/120)*x^5 + (-1/5040)*x^7
      Taylor Series (n: 3): x + (-1/6)*x^3
```

Programme 5.32 The Taylor expansion of sin(x) function with 3, 5 and 7 orders

5.3.1 Lambdify() Function in SymEngine

The ***lambdify()*** function is an important function that allows symbolic expressions to be converted into directly usable functions. In the Julia programming language, symbolic calculations can be performed with *Symbolics*, *SymEngine* or *SymPy*. Symbolic expressions constructed with these modules are in Basic or Core data structures and cannot be used directly. But ***lambdify()*** function converts the symbolic functions into Julia functions (Programme 5.34).

In Programme 5.35, first the symbolic variable $\boldsymbol{x}$ is defined and the function f is created and the value of ***f(3)*** is calculated. However, it should be noted that only $\boldsymbol{x} = \boldsymbol{3}$ is substituted instead of $\boldsymbol{x}$. The calculation of the numerical value of the function for $\boldsymbol{x} = \boldsymbol{3}$ should be transformed into a structure that Julia can interpret with the ***lambdify()*** function. This was done in cell 43 of the programme and the numerical value of the function for $\boldsymbol{x} = \boldsymbol{3}$ could be calculated (Programme 5.36).

```
[39]: using Plots

      x_values = range(-π, stop=π, length=100)
      f_values = float(f.(x_values))

      plot_legend = ["sin(x)"]
      p = plot(x_values, f_values, size=(800, 400), label="sin(x)", linewidth=2,
          color =:red, legend=:top)
      plot!(p)
```

[39]:

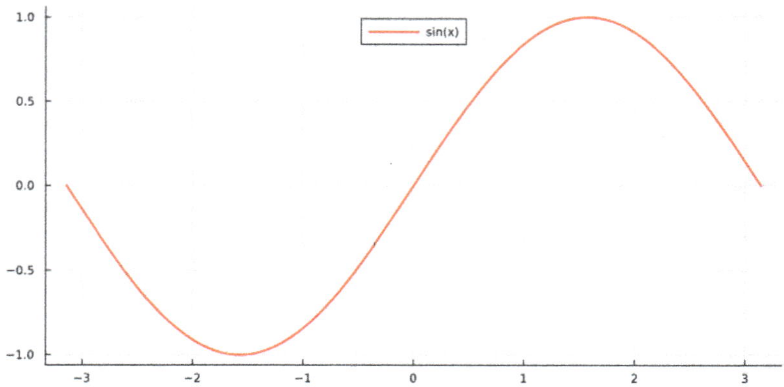

Programme 5.33 The graph of sin(x) function

5.3.2 *Derivative with SymEngine*

Differentiation is done with ***diff()*** function in this module. This function takes the name of the function and the variable for which the derivative is to be calculated as parameters. The ***lambdify()*** function mentioned above should be used for the base expressions of the derived expressions. In Programme 5.37, variable ***x*** and function ***f*** are defined with respect to this variable. The first derivative of the function f is taken with ***diff(f, x)*** and the second derivative with ***diff(diff(f, x), x)*** or ***diff(f, x, x)***.

In Programme 5.38, the velocity and acceleration of a object whose path expression is given as a function of time are calculated with ***diff(function, t)***. Then the values of path, velocity and acceleration are calculated for a given time interval. Notice that the result is returned as Basic data type. For this reason, the ***lambdify()*** function was used in Julia to obtain a more usable data set. Then, the graphs of path, velocity and acceleration were plotted (Programmes 5.39 and 5.40).

```
[40]: for (i, series_expr) in taylor_f
          taylor_fn = lambdify(series_expr, [x])

          taylor_values = taylor_fn.(x_values)

          plot!(p, x_values, taylor_values, size=(800,500), label=string("Taylor order $i"))
          push!(plot_legend, string("Taylor order $i"))
      end

      xlabel!("angel")
      ylabel!("sin(x) - Taylor expansions")
      title!("Taylor series with different orders and the sin(x) funciton")
      plot!(p)
```

[40]:

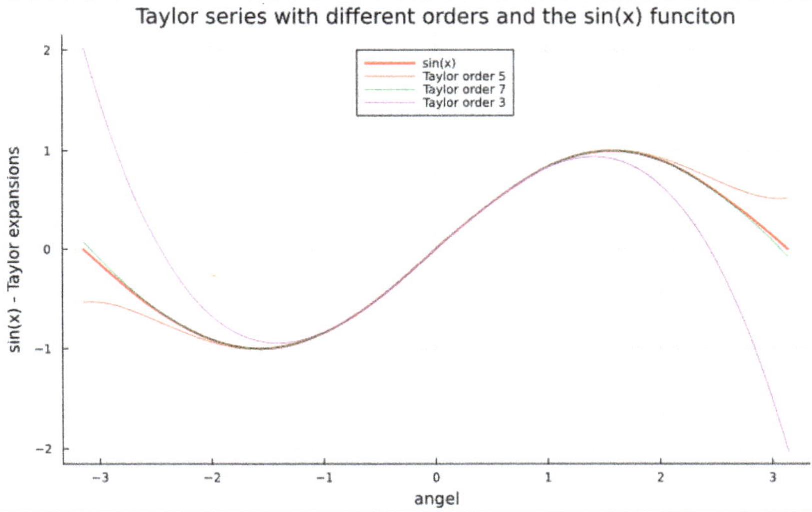

Programme 5.34 The graph of sin(x) function and its Taylor expansions

5.3.3 SymFunction() in SymEngine

So far, we have worked by defining symbolic variables using the *SymEngine* module. The module also includes symbolic function definition. This makes it possible to define functions in a more flexible way. The symbolic function is defined as given in Programme 5.41. First the name of the function is written and then in the second parenthesis the variable previously defined as a symbol.

The symbolic function defined in Programme 5.42 is defined explicitly. Then the derivative of the function is calculated.

```
[42]: x = symbols(:x)
      f = x + 3x + 2exp(-2x)
      f(3)

[42]: 12 + 2*exp(-6)

[43]: f_lamb = lambdify(f, [x])
      f_lamb(3)

[43]: 12.004957504353333

[44]: x_values = range(0, stop = 10, length = 9)

      for i in x_values
          result = f_lamb(i)
          println("for x = $i \t f(x) = $result")
      end

      for x = 0.0      f(x) = 2.0
      for x = 1.25     f(x) = 5.1641699972477975
      for x = 2.5      f(x) = 10.013475893998171
      for x = 3.75     f(x) = 15.001106168740296
      for x = 5.0      f(x) = 20.000090799859525
      for x = 6.25     f(x) = 25.000007453306345
      for x = 7.5      f(x) = 30.00000061180464
      for x = 8.75     f(x) = 35.00000005021998
      for x = 10.0     f(x) = 40.000000004122306
```

Programme 5.35 The usage of *lambdify()* function

```
[45]: α, β = symbols(:α), symbols(:β)

      f2 = cos(α) + sin(β)

      f2(30, 60)

[45]: sin(30) + cos(60)

[46]: f2_lamb = lambdify(f2, [α, β])

[46]: #36 (generic function with 1 method)

[47]: f2_lamb(pi/6, pi/3)

[47]: 1.7320508075688772
```

Programme 5.36 An another usage of *lambdify()* function

5.4 Symbolic Calculations with SymPy Module

Now we start working with the *SymPy* module, which is written for the Python programming language but can also be used with the Julia language. Since it is historically older, there are more calculations that can be done with the *Sympy* module. For example, solving equations or system of equations, limit calculation, integration and even differential equation solution. In this section, we will try to mention all the symbolic calculations that can be done with *SymPy*.

In some cases, it will be necessary to support the *SymPy* module running in Julia with *PyCall*. This is because the *SymPy* module was originally written for the Python language. The PyCa*ll* module is a module that makes it possible to use Python

```
[49]: x = symbols(:x)

      f = 7x^2 + exp(-2x)

      dfx = diff(f, x)

[49]: 14*x - 2*exp(-2*x)

[50]: dfx2 = diff(diff(f, x), x)

[50]: 14 + 4*exp(-2*x)

[51]: dfx_2 = diff(f, x, x)

[51]: 14 + 4*exp(-2*x)

[52]: dfx2(2)

[52]: 14 + 4*exp(-4)

[53]: dfx2_lamb = lambdify(dfx2, [x])
      dfx2_lamb(2)

[53]: 14.073262555554937
```

Programme 5.37 The usage of *diff()* function

```
[60]: t = symbols(:t)

      s = 3t^2 + 4t + 2
      v = diff(s, t)
      a = diff(v, t)

      println("s(t) = $s")
      println("v(t) = s'(t) = $v")
      println("a(t) = s''(t) = $a")

      s(t) = 2 + 4*t + 3*t^2
      v(t) = s'(t) = 4 + 6*t
      a(t) = s''(t) = 6
```

Programme 5.38 Velocity and acceleration of an object

modules in Julia. For example, Python modules such as numpy, pandas, matplotlib can also be run in the Julia environment.

Sym() or ***symbols()*** functions are used to define variables in *SymPy* module. ***symbols()*** function can be specified with various parameters. In addition, the use of ***symbols()*** function is similar to the symbolic variable definition features in the *SymEngine* module. For example, the parameters ***positive = true*** if the defined symbolic quantity is positive, or ***real = true*** if the defined quantity is a real number. Of course, these are special definitions because, for example, if the roots of an equation are searched and the variable is defined as a real number, the solution set will be returned as ***Any[]***. But if the variable is defined as ***real = false***, then if the roots of the equation contain complex numbers, the solution set will be obtained exactly. For this reason, the ***solve()*** function in *SymPy* is used in the following examples to show the effects of parameter definitions. This function can be used to solve equations or systems of equations (5.43).

```
[61]: t_values = range(0, stop = 1.5, length = 9)

      s_values = [subs(s, t, i) for i in t_values]
      v_values = [subs(v, t, i) for i in t_values]
      a_values = [subs(a, t, i) for i in t_values]

      println("s(t) = $s : \n", s_values)
      println("v(t) = $v : \n", v_values)
      println("a(t) = $a : \n", a_values)

      s(t) = 2 + 4*t + 3*t^2 :
      Basic[2.0, 2.85546875, 3.921875, 5.19921875, 6.6875, 8.38671875, 10.296875, 12.41796875, 14.75]
      v(t) = 4 + 6*t :
      Basic[4.0, 5.125, 6.25, 7.375, 8.5, 9.625, 10.75, 11.875, 13.0]
      a(t) = 6 :
      Basic[6, 6, 6, 6, 6, 6, 6, 6, 6]

[62]: s_lambdified = lambdify(s, [t])
      v_lambdified = lambdify(v, [t])
      a_lambdified = lambdify(a, [t])

      s_lamb_values = s_lambdified.(t_values)
      v_lamb_values = v_lambdified.(t_values)
      a_lamb_values = a_lambdified.(t_values)

      println("s = $s : ", s_lamb_values)
      println("v = $v : ", v_lamb_values)
      println("a = $a : ", a_lamb_values)

      s = 2 + 4*t + 3*t^2 : [2.0, 2.85546875, 3.921875, 5.19921875, 6.6875, 8.38671875, 10.296875, 12.41796875,
      14.75]
      v = 4 + 6*t : [4.0, 5.125, 6.25, 7.375, 8.5, 9.625, 10.75, 11.875, 13.0]
      a = 6 : [6, 6, 6, 6, 6, 6, 6, 6, 6]
```

Programme 5.39 Numerical results of the object

```
[63]: using Plots

      p = plot(t_values, s_lamb_values, size=(600, 400), label="$s", linewidth=2)
      plot!(t_values, v_lamb_values, label="$v", linewidth=2)
      plot!(t_values, a_lamb_values, label="$a", linewidth=2)

      xlabel!("t values")
      ylabel!("s and its derivatives")
      title!("Motion for a known path")

      display(p)
```

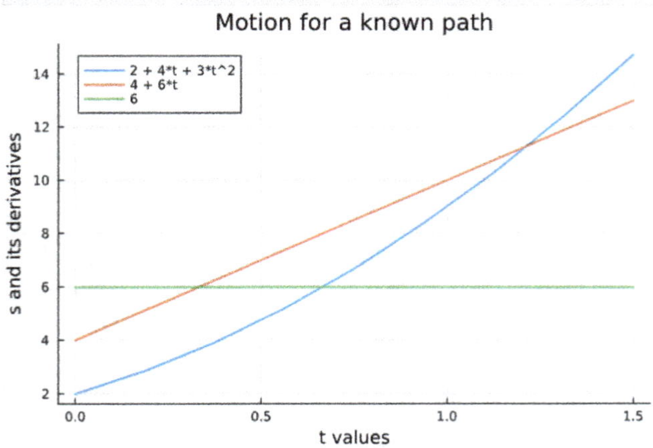

Programme 5.40 The graphics of the object

```
[65]: t = symbols(:t)
      f = SymFunction(:f)(t)
[65]: f(t)
```

Programme 5.41 Defining a symbolic function

```
[66]: f = cos(t)^2 + log(x * t)
[66]: cos(t)^2 + log(t*x)
[67]: f(t)
[67]: cos(t)^2 + log(t*x)
[68]: dft = diff(f, t)
[68]: -2*sin(t)*cos(t) + t^(-1)
[69]: dfx = diff(f, x)
[69]: x^(-1)
[70]: dft(2)
[70]: 1/2 - 2*sin(2)*cos(2)
[71]: dft_lamb = lambdify(dft, [t])
[71]: #36 (generic function with 1 method)
[72]: dft_lamb(2)
[72]: 1.2568024953079284
```

Programme 5.42 The graphics of the object

5.4.1 Solve() Function in SymPy

In the *SymPy* module, the ***solve()*** function is used to solve non-transcendental equations. Transcendental equations cannot be solved analytically and can only be solved by methods such as Newton-Raphson or Müller method. The advantage of these two methods is that if the function has imaginary roots, these methods can determine these imaginary roots (Programmes 5.44 and 5.45).

The solutions obtained by applying the function are given symbolically. However, in the following applications, these symbolic roots are also converted into numerical data that can be used in Julia language. It should be noted that if the solution of an equation is sought, it is best to define the variable without using real numbers or positive parameters. Thus, if the equation has complex roots, these roots are not missed. There is a good example of this in Programme 5.44. The expression to be solved has two roots and both are complex numbers. However, since the variable $\boldsymbol{x}$ is defined as a real number in the 12th cell of the programme, the function returns ***Any []***. In the 13th cell, since ***real = false***, the complex roots of the equation can be determined. For this reason, it will be useful to define the variable as much as

```
[2]: using SymPy

[3]: α, β, γ = Sym("α, β, γ")

[3]: (α, β, γ)

[4]: typeof(γ)

[4]: Sym{PyCall.PyObject}

[5]: x = symbols("x", real = true)

[5]: x

[6]: typeof(x)

[6]: Sym{PyCall.PyObject}

[7]: x = symbols(:x, real= true)

[7]: x

[8]: alpha = symbols("alpha")

[8]: α

[9]: typeof(alpha)

[9]: Sym{PyCall.PyObject}

[10]: beta = symbols("beta", real = false, positive = false)

[10]: β

[11]: typeof(beta)

[11]: Sym{PyCall.PyObject}
```

Programme 5.43 Defining symbolic variables in *SymPy*

possible in the solution of the equation only without specifying parameters. The ***nroots()*** function can also be used for polynomial expressions, Programme 5.48.

We can use list comprehension to print the solutions of the equation so that the roots of the equation are in a format that can be used by Julia. The roots can also

```
[12]: x = symbols(:x, real = true)
      solve(x^2 + 1)

[12]: Any[]

[13]: x = symbols(:x, real = false)
      solve(x^2 + 1)

[13]: [-i
       i]

[14]: x=symbols(:x)
      solve(x^2 + 1)

[14]: [-i
       i]
```

Programme 5.44 *solve()* function

```
[18]: beta = symbols("beta")
```

[18]: β

```
[19]: f1 = beta^3 - beta^2 - 4
```

[19]: $\beta^3 - \beta^2 - 4$

Programme 5.45 The equation for Programme 5.46

```
[20]: sol1 = solve(f1)
```

[20]: $\begin{bmatrix} 2 \\ -\frac{1}{2} - \frac{\sqrt{7}i}{2} \\ -\frac{1}{2} + \frac{\sqrt{7}i}{2} \end{bmatrix}$

```
[21]: sol1[1]
```

[21]: 2

```
[22]: sol1[2]
```

[22]: $-\frac{1}{2} - \frac{\sqrt{7}i}{2}$

```
[23]: N(sol1[2])

[23]: -0.5 - 1.3228756555322954im

[24]: typeof(N(sol1[2]))

[24]: ComplexF64 (alias for Complex{Float64})

[25]: sol1[3]
```

[25]: $-\frac{1}{2} + \frac{\sqrt{7}i}{2}$

```
[26]: N(sol1[3])

[26]: -0.5 + 1.3228756555322954im
```

Programme 5.46 The solution of $\beta^3 - \beta^2 - 4$

```
[27]: roots1 = [N(i) for i in sol1]

[27]: 3-element Vector{Number}:
         2
       -0.5 - 1.3228756555322954im
       -0.5 + 1.3228756555322954im

[28]: N.(sol1)

[28]: 3-element Vector{Number}:
         2
       -0.5 - 1.3228756555322954im
       -0.5 + 1.3228756555322954im
```

Programme 5.47 The collective results of the equation

```
[56]: f5 = x^4 + (x^3*im - 2x)^2 + 4 + (x^3-1)
```

[56]: $x^4 + x^3 + \left(ix^3 - 2x\right)^2 + 3$

```
[57]: nroots(f5)
```

[57]: $\begin{bmatrix} -1.30864857238312 + 1.12990594393196i \\ -1.12276376784796 + 0.697779685593026i \\ -0.187339126795073 - 0.778398860249697i \\ 0.203293351310884 + 0.699482819295019i \\ 0.739761099178396 - 0.966385353403304i \\ 1.67569701653688 - 0.782384235167006i \end{bmatrix}$

```
[58]: N.(nroots(f5))

[58]: 6-element Vector{ComplexF64}:
        -1.3086485723831232 + 1.1299059439319628im
        -1.122763767847961 + 0.6977796855930255im
       -0.18733912679507334 - 0.7783988602496968im
        0.20329335131088414 + 0.6994828192950187im
         0.7397610991783965 - 0.9663853534033044im
         1.6756970165368767 - 0.7823842351670057im

[59]: f5 |>string

[59]: "x^4 + x^3 + (I*x^3 - 2*x)^2 + 3"
```

Programme 5.48 The usage of *nroots()*

be converted to numerical data using the ***N()*** function. The ***N.()*** function can also be used to see them in a collective form (Programme 5.47 and 5.48).

The ***solve()*** function is also used to solve the system of equations. This usage is of two different types.

Either the equation is used in the ***solve()*** function or the names of the variables are specified in the list structure with the equation in the function. In both cases, the solution is obtained. The solution of the equation system is in the Dictionary data structure. Therefore, it consists of key value pairs. To make the results numeric, either the ***N()*** function can be used or a simple for loop can be used to convert the key-value pairs to ***Dict[key] = value.evalf()***. In Programme 5.49 a system of equations is solved. In Programme 5.50, the process of converting the solutions to numerical data is given.

5.4.2 *Simplify(), Expand(), Factor(), Together(), Apart() and .Coeff() Functions*

The *SymPy* module contains some functions for simplifying symbolic expressions and for printing expressions term by term. The ***simplify()*** function is used to simplify expressions, the ***expand()*** function is used to expand an expression term by term, the ***factor()*** function is used to factor an expression, the ***together()*** function is used to add all terms in the common denominator, the ***apart()*** function is used to separate a rational expression into its partial fractions, which are used in integral calculus or to write a rational expression in simpler fractions. In addition, the ***coeff()*** function

```
[61]: x, y = symbols("x, y")

      eqs = [4x + 3y - 2, 6x + 4y - 5]
```

[61]: $\begin{bmatrix} 4x+3y-2 \\ 6x+4y-5 \end{bmatrix}$

```
[62]: sol = solve(eqs)
```

[62]:

$$\begin{cases} y & => & -4 \\ x & => & \frac{7}{2} \end{cases}$$

```
[63]: sol = solve(eqs, [x,y])
```

[63]:

$$\begin{cases} y & => & -4 \\ x & => & \frac{7}{2} \end{cases}$$

Programme 5.49 The solution of an equation system

```
[64]: sol[x]
```

[64]: $\frac{7}{2}$

```
[65]: sol[y]
```

[65]: -4

```
[66]: println("Symbolic solutions of equation system \n: $sol")

      Symbolic solutions of equation system
      : Dict{Sym{PyCall.PyObject}, Sym{PyCall.PyObject}}(y => -4, x => 7/2)

[67]: sol_of_eq_system = Dict()
      for (key, value) in sol
          sol_of_eq_system[key] = value.evalf()
      end

[68]: println("Symbolic solutions of equation system \n: $sol_of_eq_system")

      Symbolic solutions of equation system
      : Dict{Any, Any}(y => -4.00000000000000, x => 3.50000000000000)

[69]: sol_of_eq_system[x]
```

[69]: 3.5

```
[70]: N(sol_of_eq_system[x])

[70]: 3.5

[71]: sol_of_eq_system[y]
```

[71]: -4.0

```
[72]: N(sol_of_eq_system[y])

[72]: -4.0
```

Programme 5.50 The converting the solution to the numerical data

```
[59]: f5 = x^4 + (x^3*im - 2x)^2 + 4 + (x^3-1)
```

[59]: $x^4 + x^3 + \left(ix^3 - 2x\right)^2 + 3$

```
[60]: simplify(f5)
```

[60]: $x^4 + x^3 + x^2\left(ix^2 - 2\right)^2 + 3$

```
[61]: expand(f5)
```

[61]: $-x^6 + x^4 - 4ix^4 + x^3 + 4x^2 + 3$

```
[62]: factor(f5)
```

[62]: $-x^6 - x^4\left(-1 + 4i\right) + x^3 + 4x^2 + 3$

Programme 5.51 The usage of *simplify()*, *expand()* and *factor()* functions

```
[63]: f6 = 1/(x+1) - 1/(x-1)^2 + x^4/(x^2+7)
```

[63]: $\frac{x^4}{x^2+7} + \frac{1}{x+1} - \frac{1}{(x-1)^2}$

```
[64]: together(f6)
```

[64]: $\frac{x^4(x-1)^2(x+1)+(x-1)^2(x^2+7)-(x+1)(x^2+7)}{(x-1)^2(x+1)(x^2+7)}$

```
[65]: apart(f6)
```

[65]: $x^2 - 7 + \frac{49}{x^2+7} + \frac{1}{x+1} - \frac{1}{(x-1)^2}$

```
[66]: f6.coeff(1/(x^2+7))
```

[66]: x^4

Programme 5.52 The usage of *together()*, *apart()* and *coeff()* functions

returns the coefficient of the term that is inserted into the function in an expression (Programmes 5.51, 5.52, 5.53, 5.54, 5.55 and 5.56).

The *SymPy* module is integrated into the drawing of graphs of symbolic expressions by using it in combination with the Plots module in a perfectly compatible way. In the following applications, 2D and 3D applications are given. In addition to the fact that it is a nice feature that a symbolic expression can be plotted without any special operation, it can also be plotted using the *lambdify()* function previously examined.

5.4.3 Limit, Derivative, Integrate with SymPy

With the *SymPy* module, it is possible to calculate the limit of a symbolic expression, to take the derivative and partial derivative of a symbolic expression, to solve differential equations and to calculate integrals. This section will focus on these

```
[101]: using SymPy, Plots

[102]: x = symbols("x")

[102]: x

[103]: f = x^2 - 4

[103]: x^2 - 4

[104]: plot(f, -4, 4, size=(600, 400), label = "$f")

[104]:
```

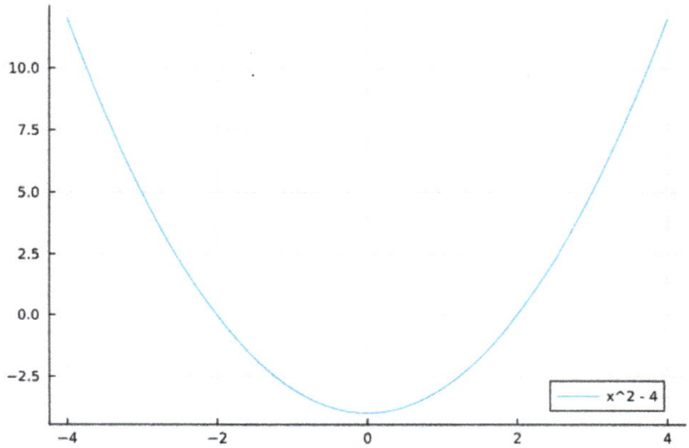

Programme 5.53 The plot of $x^2 - 4$

```
[105]: f2 = sin(x)

[105]: sin(x)

[106]: f3 = cos(x)

[106]: cos(x)

[107]: plot(sin(x), cos(x), 0, 4pi, size=(400,300), label = "$f2 & $f3", legend = :left)

[107]:
```

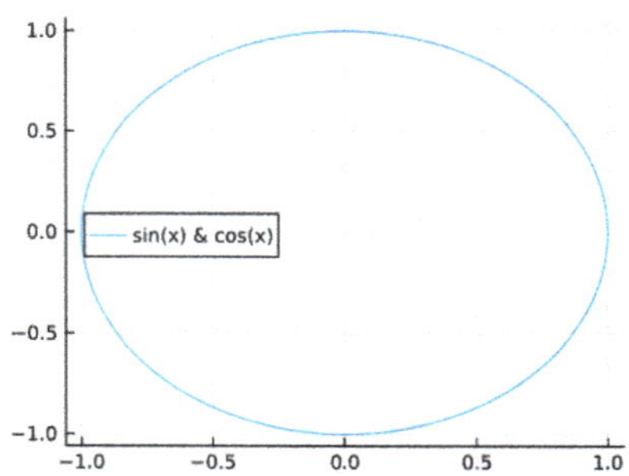

Programme 5.54 The plot of sin(x) and cos(x)

```
[111]: plot(sin(2x), cos(x), sin(x), 0, 4pi, label = "sin(2x), cos(x), sin(x)", legend = :right)
```

[111]:

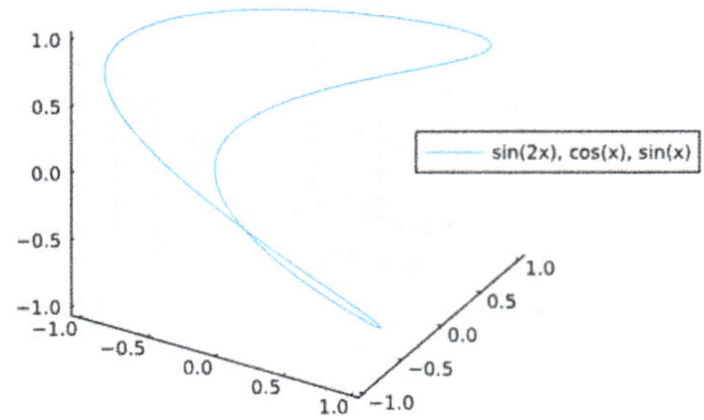

Programme 5.55 The plot of sin(2x), cos(x) and sin(x)

```
[112]: expr = sin(x)^3 + tanh(x^3)

       f7 = lambdify(expr)

       xrange = range(0, stop=10, length=256)
       plot(xrange, f7.(xrange), label = "$expr")
```

[112]:

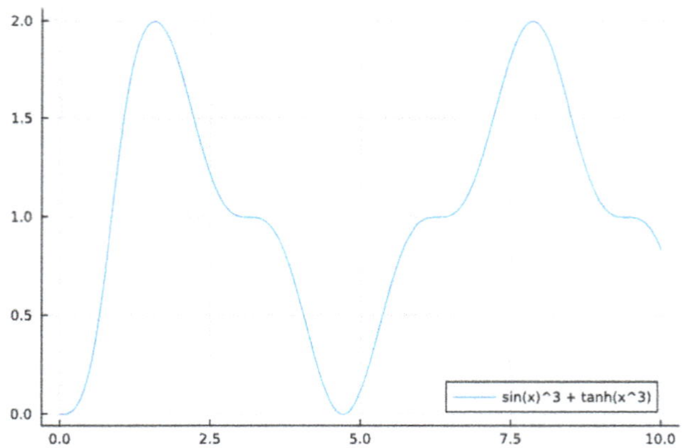

Programme 5.56 Graph of a symbolic expression as a Julia object with *lambdify()* function

calculations. For limit only, it is necessary to use ***sympy.limit()*** with the prefix *sympy* in the ***limit()*** function. Let's start with limit calculation.

```
[8]: a = symbols(:a)
```

[8]: a

```
[9]: f2 = (exp(a^3 * x^5) - log(1 + x)^4) / sqrt(2 - (a^2*x^4)^3)
```

[9]: $\frac{e^{a^3x^5} - \log(x+1)^4}{\sqrt{-a^6x^{12}+2}}$

```
[10]: f2(x => a)
```

[10]: $\frac{e^{a^8} - \log(a+1)^4}{\sqrt{2-a^{18}}}$

```
[11]: expand(f2(x => a))
```

[11]: $\frac{e^{a^8}}{\sqrt{2-a^{18}}} - \frac{\log(a+1)^4}{\sqrt{2-a^{18}}}$

```
[12]: sympy.limit(f2, x, a)
```

[12]: $\frac{e^{a^8} - \log(a+1)^4}{\sqrt{2-a^{18}}}$

```
[13]: expand(sympy.limit(f2, x, a))
```

[13]: $\frac{e^{a^8}}{\sqrt{2-a^{18}}} - \frac{\log(a+1)^4}{\sqrt{2-a^{18}}}$

Programme 5.57 The usage of *sympy.limit()* function to $\sin(x)/x$

5.4.3.1 Limit

As it is known, the convergence or divergence of an expression is extremely important. Convergence or divergence of mathematical expressions is determined by limit calculation. In Programme 5.57, the limit of the expression $\sin(x)/x$ is calculated. In the next programme, the limit of a more complex expression is given comparatively both by substituting another symbolic parameter for the variable in this complex expression and by taking the limit when directly approaching this parameter. The limit function also includes right-hand and left-hand limits (Programmes 5.58, 5.59).

5.4.3.2 Derivative

The ***diff()*** function is used for the derivative calculation. The function takes two parameters. The first parameter is the function to be differentiated, and the latter is the name of the variable.

The ***sympy.Derivative()*** function can also be used to calculate the derivative. this function takes as parameter the function to take the derivative and the variable according to which the derivative is to be taken, as in ***diff()*** function. To calculate the derivative with ***sympy.Derivative()***, the ***.doit()*** function should be used as shown in Programme 5.60.

The ***doit()*** function is a Python command and becomes operational when the *SymPy* module is included in the programme. The function evaluates symbolic expressions exactly.

```
[2]: using SymPy

[3]: x = symbols(:x)
```

[3]: x

```
[4]: f = sin(x) / x
```

[4]: $\frac{\sin(x)}{x}$

```
[5]: sympy.limit(f, x, 0)
```

[5]: 1

```
[6]: f1 = (1+1/x)^x
```

[6]: $\left(1 + \frac{1}{x}\right)^x$

```
[7]: sympy.limit(f1, x, oo)
```

[7]: e

Programme 5.58 Another application of *limit()* function

```
[15]: sympy.limit(sign(x), x, 0, dir = "-")
```

[15]: -1

```
[16]: sympy.limit(sign(x), x, 0, dir = "+")
```

[16]: 1

```
[17]: expand(sympy.limit(f2, x, a, dir = "+"))
```

[17]: $\frac{e^{a^5}}{\sqrt{2-a^{18}}} - \frac{\log(a+1)^4}{\sqrt{2-a^{18}}}$

```
[18]: expand(sympy.limit(f2, x, a, dir = "-"))
```

[18]: $\frac{e^{a^5}}{\sqrt{2-a^{18}}} - \frac{\log(a+1)^4}{\sqrt{2-a^{18}}}$

Programme 5.59 Right-hand and left-hand limits

```
[20]: x = symbols("x")
      f = exp(-x)*cos(x)
```

[20]: $e^{-x}\cos(x)$

```
[21]: dfdx = diff(f, x)
```

[21]: $-e^{-x}\sin(x) - e^{-x}\cos(x)$

```
[22]: sympy.Derivative(f, x)
```

[22]: $\frac{d}{dx}e^{-x}\cos(x)$

```
[23]: sympy.Derivative(f, x).doit()
```

[23]: $-e^{-x}\sin(x) - e^{-x}\cos(x)$

Programme 5.60 The derivative with *SymPy*

```
[25]: f
```

[25]: $e^{-x}\cos(x)$

```
[26]: dfdx = diff(f, x)
```

[26]: $-e^{-x}\sin(x) - e^{-x}\cos(x)$

```
[27]: dfdx2_ = diff(f, x, x)
```

[27]: $2e^{-x}\sin(x)$

```
[28]: dfdx2 = diff(dfdx, x)
```

[28]: $2e^{-x}\sin(x)$

```
[29]: diff(diff(f, x), x)
```

[29]: $2e^{-x}\sin(x)$

Programme 5.61 High-order derivative by applying *diff()* function

```
[31]: df = sympy.Derivative(f, x, x)
```

[31]: $\frac{d^2}{dx^2}e^{-x}\cos(x)$

```
[32]: df2 = sympy.Derivative(f, x, x).doit()
```

[32]: $2e^{-x}\sin(x)$

```
[33]: df2_ = sympy.Derivative(sympy.Derivative(f,x), x).doit()
```

[33]: $2e^{-x}\sin(x)$

Programme 5.62 High-order derivative by applying *sympy.Derivative()* function

It is possible to take higher order derivatives with ***diff()*** and ***sympy.Derivative()*** functions. For this purpose, either the derivative operation can be taken again by assigning it to another variable or the variable to be derived from is used in the function as a parameter one after the other. In Programme 5.61, high order derivative operation is performed with ***diff()*** function (Programme 5.62).

Programme 5.63 was built as a further implementation and enriched with the pipe operator (|>). In cell 35 of the programme the function ***f2*** is defined. In cell 36 the derivative operation is defined with Derivative. In cell 37, the pipe operator is used to convert the symbolic derivative into string data type. In cell 38, the derivative is calculated with the ***doit()*** function and in the next cell it is converted to string with the pipe operator. In the last cell of the programme, the solve function was applied to the expression obtained by taking the derivative with the pipe operator and the roots of the expression were obtained. Note that the solution is in *Dict* data type.

```
[35]: f2 = x^4 * a + a * x
```

[35]: $ax^4 + ax$

```
[36]: sympy.Derivative(f2, x)
```

[36]: $\frac{\partial}{\partial x}(ax^4 + ax)$

```
[37]: sympy.Derivative(f2, x) |> string

[37]: "Derivative(a*x^4 + a*x, x)"

[38]: sympy.Derivative(f2, x).doit()
```

[38]: $4ax^3 + a$

```
[39]: sympy.Derivative(f2, x).doit() |> string

[39]: "4*a*x^3 + a"

[40]: sol = sympy.Derivative(f2, x).doit() |> solve

[40]: 4-element Vector{Dict{Sym{PyCall.PyObject}, Sym{PyCall.PyObject}}}:
       Dict(a => 0)
       Dict(x => -2^(1/3)/2)
       Dict(x => 2^(1/3)/4 - 2^(1/3)*sqrt(3)*I/4)
       Dict(x => 2^(1/3)/4 + 2^(1/3)*sqrt(3)*I/4)
```

Programme 5.63 Derivation and some applications with pipe operator together

5.4.3.3 Gradient, Divergence and Curl in Julia

The derivative operations we have examined above are partial derivatives of a function. As it is known, there are derivatives such as gradient, divergence and curl which are widely used in physics. The gradient is applied to a scalar field and the result is a vector. The best example of this is that the negative value of the gradient of the potential energy in a conservative force field gives the force.

Divergence and curl, on the other hand, are applied to a force field. At the end of the divergence process, a scalar is obtained. A good example of this is that the divergence of the electric field gives the potential. Curl is also applied to a force field and the result is a new vector.

In Julia and in the *SymPy* module, special operators to perform these derivative calculations are not yet defined, but we can either write functions that perform these operations directly, or we can define them as a function for more general purposes, or we can define them as macros. It will be useful to review the third chapter of this book to define functions and macros.

Let's start with the gradient. Gradient is applied to a scalar field and in fact gradient means total differential. The proof of this will not be discussed in this book. In Programme 5.64, the gradient of a scalar field defined as ***F1*** is calculated. This is done with a simple list comprehension operation. Let's create a Julia function and ***@macro*** that calculates the gradient with the function and ***@macro*** definition rules discussed in Chap. 3. Programmes 5.65 and 5.66 show the Julia function and the ***@macro*** for calculating the gradient, respectively.

```
[49]: F1 = x^2 + x * y - x^2 * z
```

[49]: $-x^2z + x^2 + xy$

```
[50]: grad_F1 = [diff(F1, i) for i in [x, y, z]]
```

[50]: $\begin{bmatrix} -2xz+2x+y \\ x \\ -x^2 \end{bmatrix}$

```
[51]: grad_F1[1]
```

[51]: $-2xz + 2x + y$

```
[52]: grad_F1[2]
```

[52]: x

```
[53]: grad_F1[3]
```

[53]: $-x^2$

Programme 5.64 The gradient of a scaler field

```
[55]: function Gradient(f, variables)
          return [diff(f, i) for i in variables]
      end

[55]: Gradient (generic function with 1 method)

[56]: variables = [x, y, z]

      Gradient(F1, variables)
```

[56]: $\begin{bmatrix} -2xz+2x+y \\ x \\ -x^2 \end{bmatrix}$

```
[57]: Gradient(F1, [x, y, z])
```

[57]: $\begin{bmatrix} -2xz+2x+y \\ x \\ -x^2 \end{bmatrix}$

Programme 5.65 The Julia function to calculate the gradient

```
[59]: macro Grad(F, variable)
          esc(:(Gradient($F, $variable)))
      end

[59]: @Grad (macro with 1 method)

[60]: variable = [x, y, z]

      @Grad(F1, variable)
```

[60]: $\begin{bmatrix} -2xz+2x+y \\ x \\ -x^2 \end{bmatrix}$

```
[61]: @Grad(F1, [x, y, z])
```

[61]: $\begin{bmatrix} -2xz+2x+y \\ x \\ -x^2 \end{bmatrix}$

Programme 5.66 The Julia function to calculate the gradient

Now let's deal with divergence. Divergence is applied to a vector field and since the divergence itself is also a vector, it is actually like the scalar product of two vectors and the result should return a scalar value. Programme 5.67 shows the operation of the divergence operator applied to a vector field. Also in cell 65 of the same programme a Julia function called ***Div()*** has been created (Programme 5.68).

```
[63]: F2 = [x^2, x^3 * y * z, x * z]
```

[63]: $\begin{bmatrix} x^2 \\ x^3yz \\ xz \end{bmatrix}$

```
[64]: div_F = diff(F2[1], x) + diff(F2[2], y) + diff(F2[3], z)
```

[64]: $x^3z + 3x$

```
[65]: function div(f, variables)
          return sum(diff(f[i], variables[i]) for i in 1:length(variables))
      end

[65]: div (generic function with 1 method)

[66]: div(F2, [x,y, z])
```

[66]: $x^3z + 3x$

Programme 5.67 Divergence and the Julia function calculating the divergence of a vector field

```
[68]: F3 = [x^2, x^3 * y * z, x * z]
```

[68]: $\begin{bmatrix} x^2 \\ x^3yz \\ xz \end{bmatrix}$

```
[69]: curl_F = [diff(F3[3], y) - diff(F3[2], z),
                diff(F3[1], z) - diff(F3[3], x),
                diff(F3[2], x) - diff(F3[1], y)]
```

[69]: $\begin{bmatrix} -x^3y \\ -z \\ 3x^2yz \end{bmatrix}$

```
[70]: function curl(F, variables)
          Fx, Fy, Fz = F

          x, y, z = variables

          curl_x = diff(Fz, y) - diff(Fy, z)
          curl_y = diff(Fx, z) - diff(Fz, x)
          curl_z = diff(Fy, x) - diff(Fx, y)

          return [curl_x, curl_y, curl_z]
      end

[70]: curl (generic function with 1 method)

[71]: curl(F3, [x, y, z])
```

[71]: $\begin{bmatrix} -x^3y \\ -z \\ 3x^2yz \end{bmatrix}$

Programme 5.68 Curl and the Julia function calculating the curl of a vector field

```
[2]: using SymPy

[3]: x, y = symbols(:x), symbols(:y)

[3]: (x, y)

[4]: F = SymFunction("F")

[4]: F

[5]: u = diff(F(x))
```

[5]: $\frac{d}{dx}F(x)$

```
[6]: f = 3y^2 + x * y + x^2
```

[6]: $x^2 + xy + 3y^2$

```
[7]: f1 = f(y => F(x))
```

[7]: $x^2 + xF(x) + 3F^2(x)$

```
[8]: f2 = diff(f1, x)
```

[8]: $x\frac{d}{dx}F(x) + 2x + 6F(x)\frac{d}{dx}F(x) + F(x)$

```
[9]: solve(f2, u)[1]
```

[9]: $\frac{-2x-F(x)}{x+6F(x)}$

Programme 5.69 An example of implicit derivation

5.4.3.4 Implicit Derivative with Sympy

Expressions of the form $F(x, y) = 0$ are called closed functions. In order to find the derivative of such a closed function, the derivative of both sides is taken and $y' = dy/dx$ is subtracted to determine.

In Programme 5.69, first the variables $\boldsymbol{x}$ and $\boldsymbol{y}$ are defined symbolically. Then a function $\boldsymbol{F}$ is defined, and its derivative is taken in order to substitute y and to represent the duty of $\boldsymbol{y}$ to $\boldsymbol{x}$ in the derivative operation. In the next step, a function ***f*** is defined which depends on both x and y. In this expression ***dy/dx*** wants to be found. For this reason, the expression ***f1*** is formed by replacing $\boldsymbol{y}$ by ***F(x)*** in the function f and the derivative of this expression is taken with respect to $\boldsymbol{x}$. This expression was solved with solve ***dF(x)/dx*** and the desired result was found.

5.4.3.5 Integrate with SymPy

As it is known, the summation over differential intervals is defined as integral and has very important applications in both physics and engineering. For example, the work done to move another charge q from an infinite distance to a point r near a charge gives us the potential energy. To calculate this work, the integral of $dW = \vec{F} \cdot \vec{r}$ must be integrated from infinite distance to point r.

```
[2]: using SymPy

[3]: x = symbols(:x)

[3]: x

[4]: F = x^3 + x^2

[4]: x³ + x²

[5]: I = integrate(F, x)

[5]: x⁴/4 + x³/3

[6]: I |> string

[6]: "x^4/4 + x^3/3"

[7]: I(4) |> N

[7]: 256//3

[8]: I2 = integrate(F, (x, 0, 4))

[8]: 256/3
```

Programme 5.70 Indefinite integral with *integrate()* function in *SymPy*

The integral calculation is performed with the ***integrate()*** function with the *SymPy* module. As a parameter, the expression whose integral is to be calculated and the parameter indicating which variable this expression is to be taken according to are written. This is the indefinite integral. When these values are given as parameters to which variable and in which range the integral is to be calculated, the numerical value of the integral is calculated. Both the indefinite and definite integrals are given in Programme 5.70.

The ***integrate()*** function can be used for double integrals. A good example is given in Programme 5.71. The same programme also shows how to obtain the original function F by taking the derivative of the indefinitely computed integral expression. A triple integral application is given in Programme 5.72 (Programmes 5.73 and 5.74).

Symbolically, the ***sympy.Integral()*** function is used for integral representation. The parameters of the function are the same as the parameters of the ***integrate()*** function. However, in this usage, the integral is shown only symbolically. The previously mentioned ***.doit()*** function is used for the result of the integral. Both *integrate()* and *sympy.Integral()* functions can also find the results of Gamma functions.

5.4.3.6 Serial Expansions with SymPy

Series() function is a function used for series expansion of mathematical expressions. The parameters of this function are the function, according to which parameter the expression will be opened to the series, the point to be opened to the series and how many elements of the series will be taken. As known and mentioned above, the series expression of a mathematical expression around the point $\boldsymbol{x = 0}$ is a Taylor

```
[10]: x, y = symbols(:x), symbols(:y)
[10]: (x, y)
[11]: F = x^2 + x * y^2
```

[11]: $x^2 + xy^2$

```
[12]: F_integ = integrate(F, x, y)
```

[12]: $\frac{x^3 y}{3} + \frac{x^2 y^3}{6}$

```
[13]: F_integ |> string
[13]: "x^3*y/3 + x^2*y^3/6"
[14]: F_integ_diff = diff(F_integ, x, y)
```

[14]: $x\left(x + y^2\right)$

```
[15]: F_int = integrate(F, (x, 0, 3), (y, 0, 4))
```

[15]: 132

```
[16]: F_int = integrate(F, (x, 2, 3), (y, 1, 4))
```

[16]: $\frac{143}{2}$

Programme 5.71 Double integral with *integrate()* function

```
[18]: G = x^3 * y - y^2 * z^(2) + x * y * z
```

[18]: $x^3 y + xyz - y^2 z^2$

```
[19]: integrate(G, x, y, z)
```

[19]: $\frac{x^4 y^2 z}{8} + \frac{x^2 y^2 z^2}{8} - \frac{x y^3 z^3}{9}$

```
[20]: G_int = integrate(G, (x, 0, 2), (y, 1, 2), (z, 1, 2))
```

[20]: $-\frac{7}{18}$

```
[21]: float(G_int)
[21]: -0.3888888888888889
```

Programme 5.72 Triple integral with *integrate()* function

series. In Programme 5.75, Taylor series expansions of ***sin(x)***, ***1/(1 − x)*** and ***1/(1 − x²)*** functions are made. Note that the ***.remove()*** function removes terms that can be neglected.

```
[25]: integrate(sin(x^2), x)
```

[25]: $\frac{3\sqrt{2}\sqrt{\pi}S\left(\frac{\sqrt{2}x}{\sqrt{\pi}}\right)\Gamma\left(\frac{3}{4}\right)}{8\Gamma\left(\frac{7}{4}\right)}$

```
[26]: integ = sympy.Integral(sin(x^2), x)
```

[26]: $\int \sin\left(x^2\right) dx$

```
[27]: integ.doit()
```

[27]: $\frac{3\sqrt{2}\sqrt{\pi}S\left(\frac{\sqrt{2}x}{\sqrt{\pi}}\right)\Gamma\left(\frac{3}{4}\right)}{8\Gamma\left(\frac{7}{4}\right)}$

```
[28]: integ.doit() |> string
[28]: "3*sqrt(2)*sqrt(pi)*fresnels(sqrt(2)*x/sqrt(pi))*gamma(3/4)/(8*gamma(7/4))"
```

Programme 5.73 Integral with *sympy.Integral()* function and *.doit()*

```
[31]: F1 = x^2 * exp(-x^2)
```

[31]: $x^2e^{-x^2}$

```
[32]: f1_integral = sympy.Integral(F1, x)
```

[32]: $\int x^2e^{-x^2} dx$

```
[33]: f1_integral.doit()
```

[33]: $-\frac{xe^{-x^2}}{2} + \frac{\sqrt{\pi}\operatorname{erf}(x)}{4}$

```
[34]: f2_integral = sympy.Integral(F1, (x, 0, oo))
```

[34]: $\int_0^{\infty} x^2e^{-x^2} dx$

```
[35]: f2_integral.doit()
```

[35]: $\frac{\sqrt{\pi}}{4}$

```
[36]: integrate(F1, (x, 0, oo))
```

[36]: $\frac{\sqrt{\pi}}{4}$

Programme 5.74 Integral with *sympy.Integral()* function and *.doit()*

5.4.4 Solutions of Differential Equations with SymPy

With the *SymPy* module, it was previously studied that derivatives of symbolic expressions can be taken. Now in this section, we will work on solving these differential equations by creating differential equations. Before moving on to initial value problems, let us focus on solving first order and then second order differential equations.

In Programme 5.76, first the *SymPy* module is loaded and then our independent variable x is defined as a symbolic variable. When defining the symbolic function ***F***, we need to use the ***SymFunction("F")*** definition. In the 4th cell of the programme, the differential equation is defined. In this line, there are two parameters in the ***Eq()***

```
[2]: using SymPy
```

```
[3]: x = symbols(:x)
```

[3]: $$x$$

```
[4]: f1 = series(sin(x), x, 0, 7)
```

[4]: $$x - \frac{x^3}{6} + \frac{x^5}{120} + O\left(x^7\right)$$

```
[5]: f1.removeO()
```

[5]: $$\frac{x^5}{120} - \frac{x^3}{6} + x$$

```
[6]: f2 = series(1 / (1 - x), x, 0, 10)
```

[6]: $$1 + x + x^2 + x^3 + x^4 + x^5 + x^6 + x^7 + x^8 + x^9 + O\left(x^{10}\right)$$

```
[7]: f2.removeO()
```

[7]: $$x^9 + x^8 + x^7 + x^6 + x^5 + x^4 + x^3 + x^2 + x + 1$$

```
[8]: f3 = series(1 / (1 - x^2), x, 0, 10).removeO()
```

[8]: $$x^8 + x^6 + x^4 + x^2 + 1$$

```
[9]: f4 = series(log(1 / (1 - x)))
```

[9]: $$x + \frac{x^2}{2} + \frac{x^3}{3} + \frac{x^4}{4} + \frac{x^5}{5} + O\left(x^6\right)$$

```
[10]: f4.removeO()
```

[10]: $$\frac{x^5}{5} + \frac{x^4}{4} + \frac{x^3}{3} + \frac{x^2}{2} + x$$

Programme 5.75 Taylor series expansion with *SymPy*

function and the parameters are separated by commas. In fact, this spelling means that the part before the comma is the left side of the differential equation *(lhs)* and the parameter to the right of the comma is the right side of the differential equation *(rhs)*. The solution of the differential equation is performed with the function ***dsolve()***. This function also takes two parameters. The first parameter is the differential equation, and the second parameter is the parameter that specifies what the equation is to be solved according to.

In Programme 5.77, two different differential equations of the second order were solved. The first example is between the 11th and 13th cells and the second differential equation is between the 14th and 16th cells.

Let us now focus on finding the general solution of a differential equation with known initial conditions. The integral constants involved in the solution of the differential equation are determined according to the initial conditions and their numerical values are determined to obtain the exact solution of the differential equation. Let the initial condition for the differential equation be ***F(0) = 0*** for the differential equation in Programme 5.78.

In cell 18, the solution of the differential equation is recalled for recall purposes. The right-hand side of the solution is defined as ***f1***. Then, with the function ***first(free_***

```
[1]: using SymPy

[2]: x = symbols(:x)
[2]: x

[3]: F = SymFunction("F")
[3]: F

[4]: diff_eq = Eq(diff(F(x), x), x)
[4]: d/dx F(x) = x

[5]: diff_eq |> string
[5]: "Eq(Derivative(F(x), x), x)"

[6]: sol = dsolve(diff_eq, F(x))
[6]: F(x) = C1 + x^2/2

[7]: sol |> string
[7]: "Eq(F(x), C1 + x^2/2)"
```

Programme 5.76 Solution of differential equation

```
[11]: F1 = SymFunction("F1")
[11]: F1

[12]: diff_eq2 = Eq(diff(F1(x), x, x), x)
[12]: d²/dx² F1(x) = x

[13]: sol2 = dsolve(diff_eq2, F1(x))
[13]: F1(x) = C1 + C2x + x^3/6

[14]: F2 = SymFunction("F2")
[14]: F2

[15]: diff_eq3 = Eq(diff(F2(x), x, x) - x^2 * sin(x), 0)
[15]: -x^2 sin(x) + d²/dx² F2(x) = 0

[16]: sol3 = dsolve(diff_eq3, F2(x))
[16]: F2(x) = C1 - x^2 sin(x) + x(C2 - 4cos(x)) + 6sin(x)
```

Programme 5.77 Solution of differential equations

symbol(f1)), symbolic ***C1*** is assigned to the ***C1*** object from the solution of the equation. For ***f1(x = 0)*** in cell 21, ***C1*** was solved with the ***solve()*** function and the result ***[0]*** was obtained. The result of the solve function returns as a list, so the first element of this list was assigned to the ***c1_value*** object and the solution of the differential equation for this first condition was printed symbolically in cell 23 and using the ***println()*** function in cell 24.

```
[18]: diff_eq
```

[18]: $\frac{d}{dx}F(x) = x$

```
[19]: sol
```

[19]: $F(x) = C_1 + \frac{x^2}{2}$

```
[20]: f1 = sol.rhs()
```

[20]: $C_1 + \frac{x^2}{2}$

```
[21]: C1 = first(free_symbols(f1))
```

[21]: C_1

```
[22]: C1v = solve(f1(x => 0), C1)
```

[22]: [0]

```
[23]: c1_value = Sym(C1v[1])
```

[23]: 0

```
[24]: solv = sol(C1 => c1_value)
```

[24]: $F(x) = \frac{x^2}{2}$

Programme 5.78 Solution of the differential equation with the initial condition

Programme 5.79 gives the solution of a second order differential equation. It can be considered as an initial value problem for the first example given in Programme 5.77. The initial values of the problem are $\boldsymbol{F(x = 2) = 0}$ and $\boldsymbol{F'(x = 2) = 0}$. The solution of the differential equation for these initial values is given in Programme 5.77.

5.5 PyCall Module with Python Modules

The *PyCall* module is a module that allows Python code and Python modules to be used in the Julia programming language. Since Python has relatively historically more advanced modules, it is very advantageous to use modules such as *numpy*, *sympy* in the Julia environment.

After the *PyCall* module is included in the program, the desired Python module, which must have been previously installed with *Pkg*, must be added to the program by defining it as a string with the ***pyimport()*** function as in Programme 5.80. At this stage, the name of the loaded module loaded with ***pyimport()*** should be used in front of the function. ***pyimport()*** also can be included as ***@pyimport*** macro.

In the 9th cell in Programme 5.78, the *PyCall* module was first included in the programme. Then the sympy module with the abbreviation sp was included in the programme with the ***pyimport()*** function. The variable $\boldsymbol{x}$ is defined with ***sp.symbols("x")***. In the 10th cell, the function ***f1*** is defined, and in the next cell

```
[28]: sol2
```

[28]: $F_1(x) = C_1 + C_2x + \frac{x^3}{6}$

```
[29]: f2 = sol2.rhs()
```

[29]: $C_1 + C_2x + \frac{x^3}{6}$

```
[30]: C1v = solve(f2(x => 2), free_symbols(f2)[1])[1]
```

[30]: $-2C_2 - \frac{4}{3}$

```
[31]: f2 = sol2(C1 => Sym(C1v[1]))
```

[31]: $F_1(x) = C_2x - 2C_2 + \frac{x^3}{6} - \frac{4}{3}$

```
[32]: C2 = free_symbols(f2)[1]
```

[32]: C_2

```
[33]: C2v = solve(diff(f2.rhs(), x)(x => 2), free_symbols(f2)[1])
```

[33]: $[-2]$

```
[34]: solv2 = f2(C2 => Sym(C2v[1]))
```

[34]: $F_1(x) = \frac{x^3}{6} - 2x + \frac{8}{3}$

```
[35]: println("solution of $diff_eq2 : ", solv2)
      solution of Eq(Derivative(F1(x), (x, 2)), x) : Eq(F1(x), x^3/6 - 2*x + 8/3)
```

Programme 5.79 Initial value problem for second order differential equation

```
[9]: using PyCall
     sp = pyimport("sympy")

     x = sp.symbols("x")
```

[9]: x

```
[10]: f1 = sp.sin(x)^4 * sp.log(x)
```

[10]: $\log(x)\sin^4(x)$

```
[11]: expr_df = sp.diff(f1, x)
```

[11]: $4\log(x)\sin^3(x)\cos(x) + \frac{\sin^4(x)}{x}$

```
[12]: x, y = sp.symbols("x, y")

      f2 = sp.sin(x) * sp.cos(y) - sp.sin(y) * sp.cos(x)
```

[12]: $\sin(x)\cos(y) - \sin(y)\cos(x)$

```
[13]: df2_2 = sp.diff(f2, x, x)
```

[13]: $-\sin(x)\cos(y) + \sin(y)\cos(x)$

```
[14]: sp.simplify(df2_2)
```

[14]: $-\sin(x - y)$

Programme 5.80 Some algebraic calculations with *PyCall* and *sympy*

the derivative of the function ***f1*** is taken. In cell 12, the variables ***x*** and ***y*** are defined and the function ***f2*** is created with these variables and *sympy* functions. The derivative of this expression is also taken and the function is printed in a simpler way with ***sp.simplify()***.

In Programme 5.81, the indefinite integral of a function defined in cell 15 is calculated and in cell 16 the gamma function is calculated from special functions. Since the ***.evalf()*** function is used to calculate the numerical value of a symbolic expression in the sympy module in Pyrhon, the ***.evalf()*** function is used here and the numerical value of the symbolic function is calculated by defining the numerical value in the dictionary data type with subs.

In Programme 5.82, the limit of the expression ***sin(x)/x*** and in the next cell, the Taylor series expansion of the function ***exp(x)*** were made (Programme 5.83).

In the next programme, both the roots of a quadratic function were found, and the solution of a defined differential equation was performed. In Programme 5.84, the numpy module was included in the programme and the scalar product of vectors a and b was calculated.

```
[15]: f3 = sp.exp(x^2) * x
      int_f3 = sp.integrate(f3, x)
```

[15]: $\frac{e^{x^2}}{2}$

```
[16]: gamma_func = sp.functions.gamma(x)

      numerical_value = gamma_func.evalf(subs = Dict(x => 5))
```

[16]: 24.0

```
[17]: numerical_value = gamma_func.evalf(subs = Dict(x => 1/2))
```

[17]: 1.77245385090552

```
[18]: sqrt(pi)
```

```
[18]: 1.7724538509055159
```

Programme 5.81 ***sp.functions.gamma()*** and its numerical value by ***.evalf()*** function

```
[19]: f5 = sp.sin(x) / x
      limit = sp.limit(f5, x, 0)
```

[19]: 1

```
[20]: f6 = sp.exp(x)
      f6_expansion = sp.series(f6, x, 0, 7)
```

[20]: $1 + x + \frac{x^2}{2} + \frac{x^3}{6} + \frac{x^4}{24} + \frac{x^5}{120} + \frac{x^6}{720} + O\left(x^7\right)$

```
[21]: f6_expansion.removeO()
```

[21]: $\frac{x^6}{720} + \frac{x^5}{120} + \frac{x^4}{24} + \frac{x^3}{6} + \frac{x^2}{2} + x + 1$

Programme 5.82 Limit and Taylor expansion with *sympy*

```
[23]: f7 = sp.Eq(x^2 - 4x + 5, 0)
      f7_sol = sp.solve(f7, x)

[23]: 2-element Vector{PyObject}:
       PyObject 2 - I
       PyObject 2 + I

[24]: f7_sol = [PyObject(i) for i in f7_sol]

[24]: 2-element Vector{PyObject}:
       PyObject 2 - I
       PyObject 2 + I

[25]: for s in f7_sol
          println(s)
      end

      PyObject 2 - I
      PyObject 2 + I

[26]: t, k = sp.symbols("t, k")
      f = sp.Function("f")

      diffeq = sp.Eq(f(t).diff(t, t), - k * f(t))
      solution = sp.dsolve(diffeq, f(t))
```

[26]: $f(t) = C_1 e^{-t\sqrt{-k}} + C_2 e^{t\sqrt{-k}}$

Programme 5.83 The solution of an equation and solution of a differential equation

```
[30]: using PyCall
      @pyimport numpy as np

      a = np.array([1, 2, 3])
      b = np.array([4, 5, 6])
      c = np.dot(a, b)

      println("the dot product of $a and $b :", c)

      the dot product of [1, 2, 3] and [4, 5, 6] :32
```

Programme 5.84 Importing *numpy* module with *PyCall* and scalar product of two vectors

More work could be done with *PyCall*, but since it is necessary to know some Python, no further details are given here. However, it is thought that the necessary information is given for the reader who is interested in the subject, and this part will be mentioned a little more when applying it to physics problems.

5.6 Problems

1. Define a function with symbolic variables for uniformly decelerating motion. When the velocity of the object is zero, write a code that calculates the velocity of the object according to the value of the acceleration, the value of the initial velocity and the time values by preparing a loop so that the programme terminates without counting the negative velocity values.

2. Write a code that calculates the position of the object for the situation mentioned in the first question. When the velocity of the object is zero, make sure that the path taken does not return negative values.
3. Calculate the derivatives of the expressions given below using the properties of the Symbolics module.

   (a) $f_1(x) = \cos(x^2)x^3$
   (b) $f_2(x) = xe^{-x^2}$
   (c) $f_3(x) = x \ln x$
   (d) $f_4(x) = x^4 \cos x - x^2$
   (e) $f_5(x) = x^2 \ln x - x \sin^4 x$

4. Write and run Julia programmes that calculate the solution of the differential equations given below.

   (a) $\frac{dy(x)}{dx} = ax^2, \ a = \text{constant}$
   (b) $\frac{d^2y(x)}{dx^2} = ax^2, \ a = \text{constant}$
   (c) $\frac{d^2y(x)}{dx^2} = 4x^2 + 3x$
   (d) $\frac{d^2y(x)}{dx^2} = 4x^2 + 3\exp(x)$

5. Calculate the limits of the expressions given in Problem 3 and the solutions you will find in Problem 4 using Julia codes.
6. Calculate the derivatives of the expressions given in the third question using the PyCall module via the imported sympy module.
7. Calculate the integrals of the expressions given in the third question using the PyCall module via the imported sympy module.

References

1. Symbolics module. https://symbolics.juliasymbolics.org/v3.5/. Gowda, Shashi and Ma, Yingbo and Cheli, Alessandro and Gwozdz, Maja and Shah, Viral B and Edelman, Alan and Rackauckas, Christopher. High-performance symbolic-numerics via multiple dispatch, arXiv:2105.03949. 2021
2. SymEngine documentation. https://symengine.org/SymEngine.jl/. MIT License, 2015–2017 SymEngine.jl Development Team. https://github.com/symengine/SymEngine.jl
3. SymPy module. https://docs.sympy.org/latest/tutorials/intro-tutorial/index.html. Meurer A, Smith CP, Paprocki M, Čertík O, Kirpichev SB, Rocklin M, Kumar A, Ivanov S, Moore JK, Singh S, Rathnayake T, Vig S, Granger BE, Muller RP, Bonazzi F, Gupta H, Vats S, Johansson F, Pedregosa F, Curry MJ, Terrel AR, Roučka Š, Saboo A, Fernando I, Kulal S, Cimrman R, Scopatz A. 2017. SymPy: symbolic computing in Python. *PeerJ Computer Science* 3:e103 https://doi.org/10.7717/peerj-cs.103
4. PyCall module. https://github.com/JuliaPy/PyCall.jl. MIT License, 2013 by Steven G. Johnson
5. LinearAlgebra module. https://web.mit.edu/julia_v0.6.2/julia/share/doc/julia/html/en/manual/linear-algebra.html

Chapter 6
Data Frames

6.1 Introduction

Data Frames are structures used to keep and store data in a table structure and to perform various operations with the data it contains. Especially in data analysis, it is very useful to keep large data in tables. Data Frames can consist of different structures such as numbers, and textual expressions in columns. It is easy to access the data in the columns where the data is kept. It is flexible to perform operations such as grouping, filtering, and sorting on this data.

For example, the data taken in the Millikan's oil drop experiment, or any other experiment can be written to an MS Excel file and then transferred to the Julia environment as a DataFrame. Thus, many operations can be performed on this data, from manipulations to graphing the data.

In artificial intelligence and machine learning studies, data can be transferred to the Julia environment in csv (comma separated variable) or xlsx file format. In Julia, the *CSV* module [1] should be used to load csv files and the *XLSX* module [2] should be used to load xlsx files. The *DataFrames* module [3] can be used to work on the data transferred to the Julia environment. In this section, we will work on DataFrames. In addition, *PyCall* and the *pandas* [4] module for Python used in the previous section can also be used to work with DataFrames.

6.2 Importing and Exporting Files to the Julia

The *XLSX* module can be used to import an *Excel file* into the Julia environment. The ***XLSX.readxlsx("filename.xlsx")*** function is used to get information about the Microsoft Excel file before transferring it to the Julia environment. The function summarises how many datasheets the Excel file consists of and in which cells the

R. G. Türeci et al., *Julia Programming for Physics Applications*,
https://doi.org/10.1007/978-3-031-84716-5_6

```
[1]: using XLSX
     using DataFrames

[2]: data1 = XLSX.readxlsx("Excel1.xlsx")

[2]: XLSXFile("Excel1.xlsx") containing 2 Worksheets
                 sheetname size          range
     -------------------------------------------------
                    Sheet1 12x2          A1:B12
                    Sheet2 12x2          A1:B12

[3]: XLSX.sheetnames(data1)

[3]: 2-element Vector{String}:
      "Sheet1"
      "Sheet2"
```

Programme 6.1 *XLSX.readclsx()* and *XLSX.sheetnames()* functions

data is located in these datasheets. The function ***XLSX.sheetnames(filename)*** returns the names of the data sheets of the Excel file.

In Programme 6.1 the *XLSX* and *DataFrames* modules are included in the programme. The name of the file to be transferred to Julia environment is *Excel1.xlsx* file. In cell 2 of the programme, the ***XLSX.readxlsx()*** function is used to summarise the information about this file. The file contains 2 sheets. ***Sheet1*** contains data consisting of 12 × 2, i.e. 12 rows and 2 columns, and similarly ***Sheet2*** contains data consisting of 12 × 2, i.e. 12 rows and 2 columns. This data is between A1 and B12 in both data sheets. In the 3rd cell of the programme the datasheet names of the file are returned. Figure 6.1 shows a view of the MS Excel file. There are 2 datasheets and sheet1 consists of 12 rows and 2 columns.

The functions mentioned above are useful for analysing the data type properties of an MS Excel file before importing it into Julia. Now it is time to import the Excel file into Julia as a DataFrame (DF). For this purpose, it is sufficient to use the command lines executed in the 5th and 6th cells of Programme 6.2 in the ***DataFrame()*** function to display the 12 × 2 matrix structure shown as *s1data* and *s2data* as a DataFrame and pull it into the Julia environment.

Programme 6.3 represents this DF but here we have used the ***XLSX.readtable()*** function and the ***first()*** function. The first function takes two parameters. The first one is the Excel file in the same working folder, which has the extension xlsx, and the second one is the datasheet in the Excel file to be converted to DataFrame. Similarly, the ***first()*** function takes two parameters. This function takes two parameters, the first is the name of the DF and the second is an integer. This integer specifies how many first lines of the DF will be retrieved. In the first cell of Programme 6.3 the ***first(df1,3)*** definition is used to display the first 3 rows of the DF and in the next cell the first seven rows of the DF are displayed.

Now we have two DataFrames, *df1* and *df2*, which we obtained from the two datasheets of the Excel file. However, if we look at the output of the 6th and 7th cells in Programme 6.3, Any is written under the column names. This is because the type of the data extracted from the Excel file is *Any[]* by Julia. Since this data consists of decimal numbers, we must convert the data in the columns to float. The column names also contain the unit of the data in the columns. This means writing many

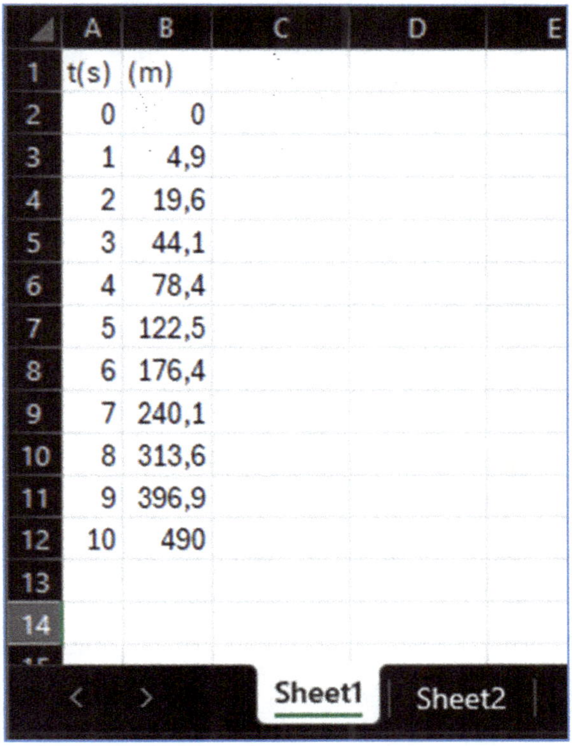

	A	B
1	t(s)	(m)
2	0	0
3	1	4,9
4	2	19,6
5	3	44,1
6	4	78,4
7	5	122,5
8	6	176,4
9	7	240,1
10	8	313,6
11	9	396,9
12	10	490

Fig. 6.1 The representation of the related Excel file

characters if we are going to perform various operations with the data in the columns. We also want to change the column names. We can change the column names first by using the ***rename!()*** function as in Programme 6.4.

Please note that *!* exclamation mark, which is entitled as "bang", is used in the rename function. This sign means that the arguments of the function are in place that is, the arguments of the function will be changed directly. Another new operator we see is the **.≥** operator. This is the broadcasting operator and the expression ***names(df1) .≥ new_column_names*** operates in such a way that each old column name is matched with the new column name. In Programme 6.6, the dataframe named *df2* was processed in the same way as in Programmes 6.4 and 6.5.

The ***show()*** function can be used to see the contents of a DataFrame more compactly. The implementation of this function is given in Programme 6.7. Notice that under the names of the columns, the type of the data in the columns is also specified as Float64. When the *eltypes* parameter is true by default in Programme 6.7, the data types of the columns are also shown. Therefore, if *eltypes* = *false*, the data type of the columns is not shown (Programme 6.8).

Now we have a DataFrame consisting of two columns showing the time and the position of the object at these moments t. With *path/time* we can determine the

```
[4]: s1data = XLSX.readdata("Excel1.xlsx", "Sheet1!A1:B12")

[4]: 12×2 Matrix{Any}:
       "t(s)"     "s(m)"
      0          0
      1           4.9
      2          19.6
      3          44.1
      4          78.4
      5         122.5
      6         176.4
      7         240.1
      8         313.6
      9         396.9
     10         490

[5]: s2data = XLSX.readdata("Excel1.xlsx", "Sheet2!A1:B12")

[5]: 12×2 Matrix{Any}:
       "t(s)"    "v(m/s)"
      0         0
      1          9.8
      2         19.6
      3         29.4
      4         39.2
      5         49
      6         58.8
      7         68.6
      8         78.4
      9         88.2
     10         98
```

Programme 6.2 *WLSX.readdata()* function to read data from every sheet in the MS Excel File

```
[6]: df1 = DataFrame(XLSX.readtable("Excel1.xlsx", "Sheet1"))
     first(df1,3)
```

[6]: 3×2 DataFrame

Row	t(s)	s(m)
	Any	Any
1	0	0
2	1	4.9
3	2	19.6

```
[7]: df2 = DataFrame(XLSX.readtable("Excel1.xlsx", "Sheet2"))
     first(df2,7)
```

[7]: 7×2 DataFrame

Row	t(s)	v(m/s)
	Any	Any
1	0	0
2	1	9.8
3	2	19.6
4	3	29.4
5	4	39.2
6	5	49
7	6	58.8

Programme 6.3 Converting DataFrame of the MS Excel File sheets

```
[9]: new_column_names = ["t", "s"]
     rename!(df1, names(df1) .=> new_column_names)

     first(df1,5)
```

[9]: 5×2 DataFrame

Row	t	s
	Any	Any
1	0	0
2	1	4.9
3	2	19.6
4	3	44.1
5	4	78.4

Programme 6.4 Changing the column names of a dataframe, bang and broadcasting operators

```
[11]: for col in names(df1)
          df1[!, col] = convert.(Float64, df1[!, col])
      end

      first(df1,5)
```

[11]: 5×2 DataFrame

Row	t	s
	Float64	Float64
1	0.0	0.0
2	1.0	4.9
3	2.0	19.6
4	3.0	44.1
5	4.0	78.4

Programme 6.5 Convert column data defined as *Any[]* of a DataFrame to float

velocity of the object at moments t. For this purpose, we need to calculate the ratio of the data in column s to the data in column t. This was done in Programme 6.9.

The *path/time* ratio was calculated with the ***df1.s./df1.t*** operation. Notice that the broadcasting operator is used again. With this operator, the same operation on a vector or a sequence of numbers is performed on each element. Dahası Note the left side of the equation. The *!* sign used in ***df1[!,:v]*** returns a reference to the original column when accessing a specific column of the DataFrame, which affects the DataFrame. In other words, the operation on the right side of the equation is added to the DataFrame as a column defined by ***:v***. As seen in Programme 6.10, column v is added to the *df1* DataFrame as a third column.

Finally, after calculating the speed, we can save the new DataFrame as an MS Excel file. The code written in Programme 6.11 can be used for this purpose. It is

```
[12]: new_column_names = ["t", "v"]
      rename!(df2, names(df2) .=> new_column_names)

      for col in names(df2)
          df2[!, col] = convert.(Float64, df2[!, col])
      end

      first(df2,5)
```

[12]: 5×2 DataFrame

Row	t	v
	Float64	Float64
1	0.0	0.0
2	1.0	9.8
3	2.0	19.6
4	3.0	29.4
5	4.0	39.2

Programme 6.6 Application of the above operations to df2 and the output of the programme

```
[13]: show(df1)

11×2 DataFrame
 Row │ t        s
     │ Float64  Float64
─────┼──────────────────
   1 │     0.0      0.0
   2 │     1.0      4.9
   3 │     2.0     19.6
   4 │     3.0     44.1
   5 │     4.0     78.4
   6 │     5.0    122.5
   7 │     6.0    176.4
   8 │     7.0    240.1
   9 │     8.0    313.6
  10 │     9.0    396.9
  11 │    10.0    490.0
```

Programme 6.7 A DataFrame by using *show()* function

easier to convert the DataFrame into a csv file and this is also provided in the same programme.

6.3 DataFrame in Julia

6.3.1 Defining a DataFrame

In Julia programming language, 3 different ways can be used to create DataFrame. The first one is to prepare an empty DataFrame and then fill this DataFrame. Programme 6.12, an empty DataFrame is created, and it is displayed as 0×0 *DataFrame*.

```
[14]: show(df1, eltypes = true)

11×2 DataFrame
 Row │ t        s
     │ Float64  Float64
─────┼──────────────────
   1 │     0.0      0.0
   2 │     1.0      4.9
   3 │     2.0     19.6
   4 │     3.0     44.1
   5 │     4.0     78.4
   6 │     5.0    122.5
   7 │     6.0    176.4
   8 │     7.0    240.1
   9 │     8.0    313.6
  10 │     9.0    396.9
  11 │    10.0    490.0

[15]: show(df1, eltypes = false)

11×2 DataFrame
 Row │ t     s
─────┼────────────
   1 │  0.0    0.0
   2 │  1.0    4.9
   3 │  2.0   19.6
   4 │  3.0   44.1
   5 │  4.0   78.4
   6 │  5.0  122.5
   7 │  6.0  176.4
   8 │  7.0  240.1
   9 │  8.0  313.6
  10 │  9.0  396.9
  11 │ 10.0  490.0
```

Programme 6.8 The *eltypes* parameter in *show()* function

```
[19]: df1[!, :v] = df1.s ./ df1.t

[19]: 11-element Vector{Float64}:
       NaN
         4.9
         9.8
        14.700000000000001
        19.6
        24.5
        29.400000000000002
        34.300000000000004
        39.2
        44.1
        49.0

[20]: first(df1, 5)

[20]: 5×3 DataFrame
```

Row	t	s	v
	Float64	Float64	Float64
1	0.0	0.0	NaN
2	1.0	4.9	4.9
3	2.0	19.6	9.8
4	3.0	44.1	14.7
5	4.0	78.4	19.6

Programme 6.9 To get velocity data and adding to the DataFrame

```
[24]: df2[!, :a] = df1.v ./ df1.t
      first(df2, 5)
```

[24]: 5×3 DataFrame

Row	t	v	a
	Float64	Float64	Float64
1	0.0	0.0	NaN
2	1.0	9.8	4.9
3	2.0	19.6	4.9
4	3.0	29.4	4.9
5	4.0	39.2	4.9

Programme 6.10 To get the acceleration data and adding to the DataFrame

save an excel file

```
[26]: filename = "myfile.xlsx"

      columns = Vector()
      push!(columns, df1[!, :t])
      push!(columns, df1[!, :s])
      push!(columns, df1[!, :v])

      labels = [ "t(s)", "s(m)", "v(m/s)"]


      XLSX.openxlsx(filename, mode="w") do df1
          sheet = df1[1]
          XLSX.writetable!(sheet, columns, labels, anchor_cell=XLSX.CellRef("B2"))
      end
```

save as a csv file

```
[28]: using CSV

      CSV.write("myfile.csv", df1)
```

[28]: "myfile.csv"

Programme 6.11 Convert a DataFrame to an MS Excel file or a CSV file

```
[1]: using DataFrames

[2]: df = DataFrame()
```

[2]: 0×0 DataFrame

Programme 6.12 Define an empty DataFrame

An empty DataFrame can be filled with different methods. The first one is Programme 6.13. The Row column specifies the row number by default. In Python it is identical to the index column. Columns *A* and *B* must be of equal length and automatically column A is defined as *integer64* and column *B* as *String*.

The second method to fill an empty DF is Programme 6.14. This method uses the ***push!()*** function, which we use to add elements to any list datatype. But only one

```
[3]: df.A = 1:3
     df.B = ["M", "F", "X"]
     df
```

[3]: 3×2 DataFrame

Row	A	B
	Int64	String
1	1	M
2	2	F
3	3	X

Programme 6.13 The first way to fill an empty DF

element can be added with push! If more than 1 element is to be added, a simple for loop must be constructed as in the second cell of the same programme. Sometimes Programme 6.15, an empty DF can be created by specifying the data types that the columns will contain.

A third way to define a DataFrame is to use the dictionary data type directly in a DataFrame. Programme 6.16, a dictionary data type named data is used directly in a DataFrame to create a *df5*.

The fourth way to define a DataFrame can perhaps be considered as a continuation of the third way. In this method, key value pairs can be transformed into a DataFrame.

```
[4]: df1 = DataFrame()
     push!(df1, (A = 4, B = 7))
```

[4]: 1×2 DataFrame

Row	A	B
	Int64	Int64
1	4	7

```
[5]: df2 = DataFrame()

     for i in 1:4
         push!(df2, (A = i, B = i + 6))
     end

     df2
```

[5]: 4×2 DataFrame

Row	A	B
	Int64	Int64
1	1	7
2	2	8
3	3	9
4	4	10

Programme 6.14 The usage of the *push!()* function to fill an empty DF

```
[6]: df3 = DataFrame( A = Int[], B = Int[] )
```

[6]: 0×2 DataFrame

Row	A	B
	Int64	Int64

```
[7]: df4 = DataFrame(A = Int[], B = String[], C = Float64[])
```

[7]: 0×3 DataFrame

Row	A	B	C
	Int64	String	Float64

Programme 6.15 Defining an empty DataFrame via indicating the columns data types

```
[8]: data = Dict("A" => [1, 2, 3], "B" => ["x", "y", "z"])
     df5 = DataFrame(data)
```

[8]: 3×2 DataFrame

Row	A	B
	Int64	String
1	1	x
2	2	y
3	3	z

Programme 6.16 Defining a DataFrame via using existing Dictionary data type

The keys are the column names, and the values are the data to be contained in the columns, Programme 6.17.

```
[9]: df6 = DataFrame(A = [1, 2, 3], B = ["x", "y", "z"])
```

[9]: 3×2 DataFrame

Row	A	B
	Int64	String
1	1	x
2	2	y
3	3	z

Programme 6.17 Defining a DataFrame via using *key-value* pairs

6.3.2 *Add a Row or a Column to an Existing DataFrame*

In order to add a line to an existing DataFrame, the ***push!()*** function can be used again. To use this function, let's start by defining a DataFrame consisting of integer numbers. Programme 6.18, a DataFrame consisting of 3×2, i.e. 3 rows and two columns, is created and a row is added to the DataFrame named *df* with the ***push!()*** function in the next cell. As a result of the operation, *df* has turned into a 4×2 DataFrame.

Let's add a column to the created DataFrame. For this operation, it is important that the number of rows is equal, otherwise an error message will be returned due to mismatch. To add columns to DataFrame, Programme 6.19, the ***insertcols!()*** function is used.

6.3.2.1 Filtering DataFrame

As previously mentioned in data mining and machine learning studies, data is transferred to the Julia environment as DataFrame. Since we are usually working with big data, analysing a DataFrame row by row and column by column is a very tedious task. However, the function ***describe(df)*** is used to make this process more compact. In addition, ***size(df)*** lists the number of rows and columns of the DataFrame, ***names(df)*** lists the column names of the DataFrame, ***eltype.(eachcol(df))*** lists the

```
[10]: df = DataFrame()
      df.A = 1:3
      df.B = ["M", "F", "X"]
      df
```

[10]: 3×2 DataFrame

Row	A Int64	B String
1	1	M
2	2	F
3	3	X

```
[11]: push!(df, (A = 4, B = "Y"))
      df
```

[11]: 4×2 DataFrame

Row	A Int64	B String
1	1	M
2	2	F
3	3	X
4	4	Y

Programme 6.18 To add a row to an existing DataFrame

```
[12]: insertcols!(df, :C => ["a1", "a2", "a3", "a4"])
      df
```

[12]: 4×3 DataFrame

Row	A	B	C
	Int64	String	String
1	1	M	a1
2	2	F	a2
3	3	X	a3
4	4	Y	a4

Programme 6.19 To add a column to an existing DataFrame

data type of each column. Programme 6.20 shows how these functions are used in Programme 6.20 shows the application of these functions to the DF defined in Programme 6.19.

Programme 6.20, the function ***describe()*** gives a table. The first column of this table is *Row*. The second column gives the column names. The third column gives the mean values of the columns, but since columns *B* and *C* are of string data type, no average value is given for B and *C*. The min column shows the smallest element in each column of *df*. Similarly, the *max* column shows the largest elements of *df*. The *median* value indicates the value of 50%. The last column shows the data type of each column. The previous column, *nmissing*, indicates whether there is missing data in each column. As can be seen, there is no missing data in columns *A*, *B* and *C*. The

```
[13]: describe(df)
```

[13]: 3×7 DataFrame

Row	variable	mean	min	median	max	nmissing	eltype
	Symbol	Union...	Any	Union...	Any	Int64	DataType
1	A	2.5	1	2.5	4	0	Int64
2	B		F		Y	0	String
3	C		a1		a4	0	String

```
[14]: size(df)

[14]: (4, 3)

[15]: names(df)

[15]: 3-element Vector{String}:
       "A"
       "B"
       "C"

[16]: eltype.(eachcol(df))

[16]: 3-element Vector{DataType}:
       Int64
       String
       String
```

Programme 6.20 The results of the usage of *describe()*, *size()*, *names()* and *eltype.(eachcol())* functions

analysis of this table contains important information for data mining and artificial intelligence studies with Julia or Python. One of the most important of these is the *nmising* column. If artificial intelligence work is to be done with the data set, there should be no missing values. In the next cells of the same programme, the size of the DF is given as (4, 3) and is the result of the ***size()*** function. ***names(df)*** show the column names of the DataFrame, and the next cell shows the data type of each column.

6.3.3 Missing Values in a DataFrame

In the previous section we focused on obtaining general information about a DataFrame by using various functions on a DataFrame. Now let us define a new DataFrame with missing data and try to find the missing data. This new DataFrame is defined using key value pairs, Programme 6.21. This new DataFrame has a size of 5×2, i.e. 5 rows and 2 columns. Note that there are 2 missing values in column A and 2 missing values in column B. With the ***describe(df)*** function, this result is also confirmed in the *nmissing* column. It should be noted that if we are working with a *df* containing thousands of data, then the ***describe()*** function is very useful. On the other hand, the data type of columns A and B is *int64?*, so missing data should be suspected.

Programme 6.22, with the function *any(ismissing.(df.A))*, it is queried whether there is a missing value in DF and the answer is *true*. In the next cell, *ismissing.(df.A)*

```
[17]: df = DataFrame(A = [1, missing, 3, 4, missing], B = [4, 5, missing, missing, 8])
      df
```

[17]: 5×2 DataFrame

Row	A	B
	Int64?	Int64?
1	1	4
2	*missing*	5
3	3	*missing*
4	4	*missing*
5	*missing*	8

```
[18]: describe(df)
```

[18]: 2×7 DataFrame

Row	variable	mean	min	median	max	nmissing	eltype
	Symbol	Float64	Int64	Float64	Int64	Int64	Union
1	A	2.66667	1	3.0	4	2	Union{Missing, Int64}
2	B	5.66667	4	5.0	8	2	Union{Missing, Int64}

Programme 6.21 A DataFrame with missing values

```
[19]: any(ismissing.(df.A))

[19]: true

[20]: ismissing.(df.A)

[20]: 5-element BitVector:
       0
       1
       0
       0
       1

[21]: ismissing.(df.B)

[21]: 5-element BitVector:
       0
       0
       1
       1
       0
```

Programme 6.22 Determination of missing values

```
[22]: missing_indices_A = findall(ismissing, df.A)

[22]: 2-element Vector{Int64}:
       2
       5

[23]: missing_indices_B = findall(ismissing, df.B)

[23]: 2-element Vector{Int64}:
       3
       4
```

Programme 6.23 Determination of index or row number of missing values

and *ismissing.(df.B)* are used to ask for missing values in columns A and B. The result is given as a column and *0* means *false* and *1* means *true*. By using the *ismissing* statement in Programme 6.22 in the ***findall()*** function as in Programme 6.23, we can see the index or row numbers of the missing data more clearly.

In artificial intelligence studies, missing values are usually replaced by the mean value of the data column so that the data set is not corrupted. However, these topics are beyond the scope of this book.

6.3.4 Indexing of a DataFrame

It may be necessary to select only a particular column or only a particular row or only an element or elements in a particular row of a particular column of any DataFrame. We call this indexing, and it is performed with the properties specified in this section. Let's clarify this issue by defining a new DataFrame. This new DataFrame is given in Programme 6.24.

```
[24]: df = DataFrame(Number = 1:10,
          String = ["One", "Two", "Three", "Four", "Five", "Six", "Seven", "Eight", "Nine", "Ten"])

[24]: 10×2 DataFrame
```

Row	Number	String
	Int64	String
1	1	One
2	2	Two
3	3	Three
4	4	Four
5	5	Five
6	6	Six
7	7	Seven
8	8	Eight
9	9	Nine
10	10	Ten

Programme 6.24 A new df

To select all rows in the Number column of the DataFrame named df, the method as in Programme 6.25 can be used. In this method, the command line ***df[:,:Number]*** is used.

Another way to select a column as it is to use the command line ***df. "Number"*** as shown in Programme 6.26.

```
[25]: df[:, :Number]

[25]: 10-element Vector{Int64}:
        1
        2
        3
        4
        5
        6
        7
        8
        9
       10
```

Programme 6.25 Selection of all Number column

```
[26]: df."Number"

[26]: 10-element Vector{Int64}:
        1
        2
        3
        4
        5
        6
        7
        8
        9
       10
```

Programme 6.26 Another way to select a whole column

```
[27]: df[4, :]
```

[27]: DataFrameRow (2 columns)

Row	Number	String
	Int64	String
4	4	Four

Programme 6.27 To select a certain row in a DataFrame

So it is clear that the first expression before the comma specifies the row and the second data after the comma specifies the column. In this case, the expression ***df[:,:Number]*** returns all rows of the Number column. Similarly, if ***df[4,:]*** notation takes all columns coming to the fourth row (Programme 6.27).

6.4 Some Applications of DataFrames

The function ***f* = *sin(x)*** defined by the variable x defined symbolically in Programme 6.28 is defined below. Then, with ***lambdify***, the function s is transformed by Julia into a function that returns a numeric value. From 0 to 360°, a degree list named D was prepared with 15° intervals. Since trigonometric functions work in radians, the D list was converted to radian by using the ***deg2rad()*** function in the ***map()*** function.

In the next step, an empty DataFrame was created by adding the DataFrame module to the programme with using *DataFrames*. Then a column named *df.theta_*

```
[1]: using SymPy

[2]: x = symbols(:x)
```

[2]: x

```
[3]: f = sin(x)
```

[3]: $\sin(x)$

```
[4]: s = lambdify(f)

[4]: #152 (generic function with 1 method)

[5]: D = [i for i in 0:15:360]
     radian = map(deg2rad, D)
     S = map(s, radian)

     using DataFrames

     df = DataFrame()
```

[5]: 0×0 DataFrame

Programme 6.28 Defining a DataFrame for trigonometric function values

```
[6]: df.teta_deg = D
     df.teta_rad = radian
     df.sin = S

     first(df, 5)
```

[6]: 5×3 DataFrame

Row	teta_deg	teta_rad	sin
	Int64	Float64	Float64
1	0	0.0	0.0
2	15	0.261799	0.258819
3	30	0.523599	0.5
4	45	0.785398	0.707107
5	60	1.0472	0.866025

Programme 6.29 First 5 rows of the DataFrame

```
[7]: last(df, 5)
```

[7]: 5×3 DataFrame

Row	teta_deg	teta_rad	sin
	Int64	Float64	Float64
1	300	5.23599	-0.866025
2	315	5.49779	-0.707107
3	330	5.75959	-0.5
4	345	6.02139	-0.258819
5	360	6.28319	-2.44929e-16

```
[8]: last_rows = df[end-4:end, :]
```

[8]: 5×3 DataFrame

Row	teta_deg	teta_rad	sin
	Int64	Float64	Float64
1	300	5.23599	-0.866025
2	315	5.49779	-0.707107
3	330	5.75959	-0.5
4	345	6.02139	-0.258819
5	360	6.28319	-2.44929e-16

Programme 6.30 Last 5 rows of the DataFrame by two different definitions

deg is created with the elements of the D list and a column named *df.theta_rad* is created with the elements of the radian list. A third column was created with the sine function values transformed with the ***lambdify()*** function (Programmes 6.29 and 6.30).

After the DataFrame was created, the first 5 lines of the DataFrame were printed with ***first(df, 5)*** and the last 5 lines of the DataFrame were printed with ***last(df, 5)***.

```
[9]: println("first 5 rows:")
     println(first(df, 5))
     println("\n")
     println("last five rows:")
     println(last(df, 5))

     first 5 rows:
     5×3 DataFrame
      Row │ teta_deg  teta_rad  sin
          │ Int64     Float64   Float64
     ─────┼──────────────────────────────
        1 │        0  0.0       0.0
        2 │       15  0.261799  0.258819
        3 │       30  0.523599  0.5
        4 │       45  0.785398  0.707107
        5 │       60  1.0472    0.866025
```

Programme 6.31 The representation of a DataFrame with *println()* function

The last five rows of the DataFrame can also be displayed with the command line ***last_rows = df [end-4: end,:]***. In the next step, the first five and last five lines of the DataFrame are printed with the ***println()*** function, but the difference in display should be noted (Programme 6.31).

Programme 6.32 shows the entire DataFrame with the function ***show(df, eltypes = false)***. As already mentioned, *eltypes* indicates by default the type of data contained in the columns of the DataFrame. Setting this parameter to false hides the column data types. It is important to note that the bottom line of the DataFrame display contains a warning in blue: "*3 rows omitted*". This warning means that 3 rows are omitted.

In the next cell the functions ***cos()***, ***sin(x)²*** + ***cos(x)²***, ***2sin(x)cos(x)***, ***sin(2x)*** are added directly to the DataFrame named df with list comprehension. Then ***df.tan*** column was added to the DataFrame with ***df.sin./df.cos***. As mentioned above, the symbol. is the broadcasting operator. Finally, the DataFrame is displayed with the ***show()*** function. This time the warning "*2 columns and 3 rows omitted*" is displayed at the bottom of the DataFrame (Programme 6.33).

The way to make these hidden columns and rows visible may be possible by changing the settings of the ***show()*** function. For this purpose, import ***Base.show*** to import the show function from Julia's Base module. The settings of the ***show()*** function can be changed by following the steps in Programme 6.34. In this cell, we can explain the parameters in ***Base.show()*** as follows. ***io::IO*** specifies the input and output parameter, that is, where the output will be made. The ***df::DataFrame*** parameter specifies for which object the ***show()*** function is updated. These parameters are synchronised to the new ***show()*** function settings. setting ***allrows = true*** completes the setting of the show function to show all rows and columns of the DataFrame.

In Julia, *DataFrame* objects can also be displayed as tables with the *PrettyTables* module [5]. For this purpose, after the PrettyTables module is included in the programme, the DataFrame can be displayed as a table with the command line ***pretty_table(DataFrame name)***. This application is given in Programme 6.35. Finally, with the Plots module, all columns of this DataFrame, that is, the numerical values of the

```
[10]: show(df, eltypes = false)

25×3 DataFrame
 Row │ teta_deg  teta_rad  sin
─────┼──────────────────────────────────
   1 │        0  0.0       0.0
   2 │       15  0.261799  0.258819
   3 │       30  0.523599  0.5
   4 │       45  0.785398  0.707107
   5 │       60  1.0472    0.866025
   6 │       75  1.309     0.965926
   7 │       90  1.5708    1.0
   8 │      105  1.8326    0.965926
   9 │      120  2.0944    0.866025
  10 │      135  2.35619   0.707107
  11 │      150  2.61799   0.5
  ⋮  │    ⋮         ⋮          ⋮
  15 │      210  3.66519   -0.5
  16 │      225  3.92699   -0.707107
  17 │      240  4.18879   -0.866025
  18 │      255  4.45059   -0.965926
  19 │      270  4.71239   -1.0
  20 │      285  4.97419   -0.965926
  21 │      300  5.23599   -0.866025
  22 │      315  5.49779   -0.707107
  23 │      330  5.75959   -0.5
  24 │      345  6.02139   -0.258819
  25 │      360  6.28319   -2.44929e-16
                              3 rows omitted
```

Programme 6.32 The representation of whole DataFrame

```
[11]: df.cos = [cos(i) for i in radian]
      df.sin2sos2 = [sin(i)^2 + cos(i)^2 for i in S]
      df.sin2 = [2 * sin(i) * cos(i) for i in S]
      df.sin2t = [sin(2 * i) for i in S]
      df[!, :tan] = df.sin ./ df.cos

      show(df, eltypes = false)

25×8 DataFrame
 Row │ teta_deg  teta_rad  sin           cos           sin2sos2  sin2          …
─────┼──────────────────────────────────────────────────────────────────────────
   1 │        0  0.0       0.0           1.0                1.0  0.0           …
   2 │       15  0.261799  0.258819      0.965926           1.0  0.494829
   3 │       30  0.523599  0.5           0.866025           1.0  0.841471
   4 │       45  0.785398  0.707107      0.707107           1.0  0.987766
   5 │       60  1.0472    0.866025      0.5                1.0  0.987027      …
   6 │       75  1.309     0.965926      0.258819           1.0  0.935525
   7 │       90  1.5708    1.0           6.12323e-17        1.0  0.909297
   8 │      105  1.8326    0.965926      -0.258819          1.0  0.935525
   9 │      120  2.0944    0.866025      -0.5               1.0  0.987027      …
  10 │      135  2.35619   0.707107      -0.707107          1.0  0.987766
  11 │      150  2.61799   0.5           -0.866025          1.0  0.841471
  ⋮  │    ⋮         ⋮          ⋮             ⋮            ⋮          ⋮          ⋱
  15 │      210  3.66519   -0.5          -0.866025          1.0  -0.841471
  16 │      225  3.92699   -0.707107     -0.707107          1.0  -0.987766     …
  17 │      240  4.18879   -0.866025     -0.5               1.0  -0.987027
  18 │      255  4.45059   -0.965926     -0.258819          1.0  -0.935525
  19 │      270  4.71239   -1.0          -1.83697e-16       1.0  -0.909297
  20 │      285  4.97419   -0.965926     0.258819           1.0  -0.935525     …
  21 │      300  5.23599   -0.866025     0.5                1.0  -0.987027
  22 │      315  5.49779   -0.707107     0.707107           1.0  -0.987766
  23 │      330  5.75959   -0.5          0.866025           1.0  -0.841471
  24 │      345  6.02139   -0.258819     0.965926           1.0  -0.494829     …
  25 │      360  6.28319   -2.44929e-16  1.0                1.0  -4.89859e-16
                                                   2 columns and 3 rows omitted
```

Programme 6.33 2 columns and 2 rows omitted in DataFrame

```
[12]: import Base.show

      Base.show(io::IO, df::DataFrame) = show(io, df, allrows = true)

      println(df)

      25×8 DataFrame
       Row │ teta_deg  teta_rad  sin           cos           sin2sos2  sin2          sin2t         tan
           │ Int64     Float64   Float64       Float64       Float64   Float64       Float64       Float64
      ─────┼─────────────────────────────────────────────────────────────────────────────────────────────────
         1 │        0  0.0       0.0           1.0                1.0  0.0           0.0           0.0
         2 │       15  0.261799  0.258819      0.965926           1.0  0.494829      0.494829      0.267949
         3 │       30  0.523599  0.5           0.866025           1.0  0.841471      0.841471      0.57735
         4 │       45  0.785398  0.707107      0.707107           1.0  0.987766      0.987766      1.0
         5 │       60  1.0472    0.866025      0.5                1.0  0.987027      0.987027      1.73205
         6 │       75  1.309     0.965926      0.258819           1.0  0.935525      0.935525      3.73205
         7 │       90  1.5708    1.0           6.12323e-17        1.0  0.909297      0.909297      1.63312e16
         8 │      105  1.8326    0.965926     -0.258819           1.0  0.935525      0.935525     -3.73205
         9 │      120  2.0944    0.866025     -0.5                1.0  0.987027      0.987027     -1.73205
        10 │      135  2.35619   0.707107     -0.707107           1.0  0.987766      0.987766     -1.0
        11 │      150  2.61799   0.5          -0.866025           1.0  0.841471      0.841471     -0.57735
        12 │      165  2.87979   0.258819     -0.965926           1.0  0.494829      0.494829     -0.267949
        13 │      180  3.14159   1.22465e-16  -1.0                1.0  2.44929e-16   2.44929e-16  -1.22465e-16
        14 │      195  3.40339  -0.258819     -0.965926           1.0 -0.494829     -0.494829      0.267949
        15 │      210  3.66519  -0.5          -0.866025           1.0 -0.841471     -0.841471      0.57735
```

Programme 6.34 Changing the settings of the show function for DataFrame object for only certain rows

```
[13]: using PrettyTables

[14]: pretty_table(df)
```

teta_deg Int64	teta_rad Float64	sin Float64	cos Float64	sin2sos2 Float64	sin2 Float64	sin2t Float64	tan Float64
0	0.0	0.0	1.0	1.0	0.0	0.0	0.0
15	0.261799	0.258819	0.965926	1.0	0.494829	0.494829	0.267949
30	0.523599	0.5	0.866025	1.0	0.841471	0.841471	0.57735
45	0.785398	0.707107	0.707107	1.0	0.987766	0.987766	1.0
60	1.0472	0.866025	0.5	1.0	0.987027	0.987027	1.73205
75	1.309	0.965926	0.258819	1.0	0.935525	0.935525	3.73205
90	1.5708	1.0	6.12323e-17	1.0	0.909297	0.909297	1.63312e16
105	1.8326	0.965926	-0.258819	1.0	0.935525	0.935525	-3.73205
120	2.0944	0.866025	-0.5	1.0	0.987027	0.987027	-1.73205
135	2.35619	0.707107	-0.707107	1.0	0.987766	0.987766	-1.0
150	2.61799	0.5	-0.866025	1.0	0.841471	0.841471	-0.57735
165	2.87979	0.258819	-0.965926	1.0	0.494829	0.494829	-0.267949
180	3.14159	1.22465e-16	-1.0	1.0	2.44929e-16	2.44929e-16	-1.22465e-16
195	3.40339	-0.258819	-0.965926	1.0	-0.494829	-0.494829	0.267949
210	3.66519	-0.5	-0.866025	1.0	-0.841471	-0.841471	0.57735

Programme 6.35 *pretty_table(df)* representation of the DataFrame for only certain rows

functions of interest above, can be plotted graphically according to the angle values (Programme 6.36).

Now let's create a DataFrame through another example. Let's define the function $f = 1/\log(x + 1)$ by defining the variable x as a symbolic parameter using the *SymPy* module. Then let's create Taylor series expansions of the first 5, first 7 and first 9 terms of this function. Then, let's create a list of independent variables from 0 to 1 with a step interval of 0.1 and compare the values of Taylor series expansions for each x value.

```
[17]: using Plots

[18]: p = plot(df.teta_deg, df.sin, label = "sin(x)",
          title = "Some trigonometric function values",
          xlabel = "θ in degrees", ylabel = "trigonometric functions")
      plot!(df.teta_deg, df.cos, label = "cos(x)")
      plot!(df.teta_deg, df.sin2sos2, label = "sin(x)^2 + cos(x)^2")
      plot!(df.teta_deg, df.sin2, label = "2 sin(x) cos(x)")

[18]:
```

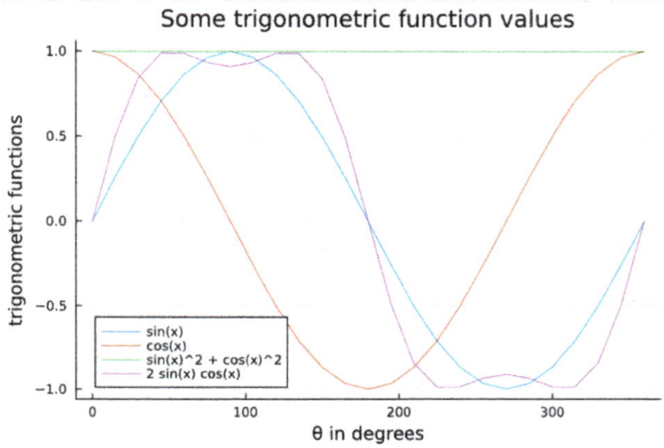

Programme 6.36 The graphics of the columns of the DataFrame for varying angle values

According to the results obtained, the Taylor series expansion consisting of 9 terms gave the same result as the value of the function for the first five steps (Programmes 6.37, 6.38, 6.39 and 6.40).

```
[1]: using SymPy

[2]: x = symbols(:x)
```

[2]: x

```
[3]: f = 1 / (log(x+1))
```

[3]: $\frac{1}{\log(x+1)}$

```
[4]: f1 = series(f, x, 0, 5).removeO()
```

[4]: $\frac{3x^4}{160} - \frac{19x^3}{720} + \frac{x^2}{24} - \frac{x}{12} + \frac{1}{2} + \frac{1}{x}$

```
[5]: f2 = series(f, x, 0, 7).removeO()
```

[5]: $\frac{275x^6}{24192} - \frac{863x^5}{60480} + \frac{3x^4}{160} - \frac{19x^3}{720} + \frac{x^2}{24} - \frac{x}{12} + \frac{1}{2} + \frac{1}{x}$

```
[6]: f3 = series(f, x, 0, 9).removeO()
```

[6]: $\frac{8183x^8}{1036800} - \frac{33953x^7}{3628800} + \frac{275x^6}{24192} - \frac{863x^5}{60480} + \frac{3x^4}{160} - \frac{19x^3}{720} + \frac{x^2}{24} - \frac{x}{12} + \frac{1}{2} + \frac{1}{x}$

Programme 6.37 The Taylor expansion of $1/\log(x + 1)$ function

```
[7]: xx = [i for i in 0:0.1:1]

     fx = lambdify(f)
     f1x = lambdify(f1)
     f2x = lambdify(f2)
     f3x = lambdify(f3)

[7]: #152 (generic function with 1 method)

[8]: F = map(fx, xx)
     F1 = map(f1x, xx)
     F2 = map(f2x, xx)
     F3 = map(f3x, xx)

     using DataFrames

     df = DataFrame()

[8]: 0×0 DataFrame
```

Programme 6.38 The numerical results of the function and its serial expansions

```
[9]: df.x = xx
     df.f = F
     df.F1 = F1
     df.F2 = F2
     df.F3 = F3

     first(df,10)

[9]: 10×5 DataFrame
```

Row	x	f	F1	F2	F3
	Float64	Float64	Float64	Float64	Float64
1	0.0	Inf	Inf	Inf	Inf
2	0.1	10.4921	10.4921	10.4921	10.4921
3	0.2	5.48481	5.48482	5.48482	5.48481
4	0.3	3.81149	3.81152	3.8115	3.81149
5	0.4	2.97201	2.97212	2.97202	2.97201
6	0.5	2.4663	2.46662	2.46635	2.46631
7	0.6	2.12764	2.1284	2.12782	2.12769
8	0.7	1.88456	1.88611	1.88504	1.88473
9	0.8	1.7013	1.70417	1.70247	1.70184
10	0.9	1.55799	1.56293	1.56054	1.55946

Programme 6.39 The DataFrame which contains the numerical results of the function and its serial expansions

6.5 Problems

1. Create a DataFrame containing the position and velocity information of an object released freely from the summit of Mount Everest when it reaches sea level and accept the acceleration of gravity as constant.

```
[10]: using PrettyTables

[11]: pretty_table(df)

┌─────────┬─────────┬─────────┬─────────┬─────────┐
│       x │       f │      F1 │      F2 │      F3 │
│ Float64 │ Float64 │ Float64 │ Float64 │ Float64 │
├─────────┼─────────┼─────────┼─────────┼─────────┤
│     0.0 │     Inf │     Inf │     Inf │     Inf │
│     0.1 │ 10.4921 │ 10.4921 │ 10.4921 │ 10.4921 │
│     0.2 │ 5.48481 │ 5.48482 │ 5.48482 │ 5.48481 │
│     0.3 │ 3.81149 │ 3.81152 │  3.8115 │ 3.81149 │
│     0.4 │ 2.97201 │ 2.97212 │ 2.97202 │ 2.97201 │
│     0.5 │  2.4663 │ 2.46662 │ 2.46635 │ 2.46631 │
│     0.6 │ 2.12764 │  2.1284 │ 2.12782 │ 2.12769 │
│     0.7 │ 1.88456 │ 1.88611 │ 1.88504 │ 1.88473 │
│     0.8 │  1.7013 │ 1.70417 │ 1.70247 │ 1.70184 │
│     0.9 │ 1.55799 │ 1.56293 │ 1.56054 │ 1.55946 │
│     1.0 │  1.4427 │ 1.45069 │ 1.44779 │ 1.44633 │
└─────────┴─────────┴─────────┴─────────┴─────────┘
```

Programme 6.40 The *pretty_table* representation of the DataFrame

2. Can you repeat the previous question as a function of the radius of the earth and the height above sea level, assuming that the acceleration of gravity is not constant, i.e. the earth is a perfectly smooth sphere? Note that this time the acceleration of gravity cannot be constant.
3. Record the data of the current passing through this resistor according to the changing voltage applied to the two ends of an R resistor in an MS Excel file and try to transfer this file to the Julia environment and determine the value of the R resistor.
4. Imagine that you conduct an experiment using the same metal and different frequencies of light to observe the photoelectric effect. Assume that from the moment the ammeter starts to show the current value, you save the values you have determined in an MS Excel file and transfer them to the Julia environment. Can you determine the kinetic energy of electrons, and the work function of the metal based on the data contained in such a DataFrame. The stage is yours.
5. Imagine repeating the previous question for 5 different metals, each with 10 measurements, and transferring all the data to the Julia environment as a single DataFrame. In order to determine the work function for each metal, you will need to work selectively on certain rows. How can you do these operations?

References

1. CSV module. https://github.com/JuliaData/CSV.jl; MIT "Expat" License. 2015: Jacob Quinn, JuliaData contributors, Julia Computing
2. XLSX module. https://felipenoris.github.io/XLSX.jl/stable/; MIT License. 2018–2023 Felipe Noronha Tavaresand other contributors: https://github.com/felipenoris/XLSX.jl/contributors
3. DataFrames module. https://dataframes.juliadata.org/stable/; MIT License. 2012–2022: Harlan Harris, EPRI (Tom Short's code), Chris DuBois, John Myles White, Milan Bouchet-Valat, Bogumił Kamiński and other DataFrames.jl contributors

4. pandas module. https://pandas.pydata.org/; 2008, *pandas* development began at https://www.aqr.com/; 2015, https://numfocus.org/sponsored-projects
5. PrettyTables module. https://ronisbr.github.io/PrettyTables.jl/stable/; MIT License. 2019–2024: Ronan Arraes Jardim Chagas

Part II
Physics Applications with Julia Codes

Chapter 7
Julia Programming Applications in Mechanics

7.1 Introduction

Mechanical physics has a wide range of applications across various fields. Engineering, aerospace, biomechanics, robotics, energy systems and earth sciences are some key areas where it plays a crucial role.

In this section, applications related to mechanics, which is one of the very large fields of physics, will be discussed. For the problems that we will encounter in this field, program codes and references will be given using the Julia programming language.

The approach to physics problems will be from simple to complex. Thus, a clearer understanding of the programming algorithm will be provided.

7.2 Scalar and Vector Product of Vectors

Vectors are physical quantities that have both magnitude and direction. They can be represented in various dimensions but are commonly represented in two or three dimensions. A vector in two dimensions $\vec{A} = A_x\hat{i} + A_y\hat{j}$, a vector in three dimensions $\vec{A} = A_x\hat{i} + A_y\hat{j} + A_z\hat{k}$ can be expressed in the forms.

In general, the scalar product of any two vectors $\vec{A}$ and $\vec{B}$ is a scalar quantity equal to the product of the magnitudes of the two vectors and the cosine of the angle θ between them:

$$\vec{A} \cdot \vec{B} = \left|\vec{A}\right|\left|\vec{B}\right|\cos\theta \tag{7.1}$$

$\left|\vec{B}\right|\cos\theta$ is the projection of $\vec{B}$ onto $\vec{A}$. Therefore, Eq. 7.1 says that $\left|\vec{A}\right|\left|\vec{B}\right|$ is the product of the magnitude of $\vec{A}$ and the projection of $\vec{B}$ onto $\vec{A}$.

R. G. Türeci et al., *Julia Programming for Physics Applications*,
https://doi.org/10.1007/978-3-031-84716-5_7

The unit vectors $\hat{\imath}, \hat{\jmath}$ and $\hat{k}$, lie in the positive x, y, and z directions, respectively, of a right-handed coordinate system. The magnitude of each unit vector equals 1. Therefore, it follows from the definition of $\vec{A} \cdot \vec{B}$ that the scalar products of these unit vectors are

$$\vec{A} = A_x\hat{\imath} + A_y\hat{\jmath} + A_z \cdot \hat{k}, \tag{7.2}$$

$$\vec{B} = B_x\hat{\imath} + B_y\hat{\jmath} + B_z \cdot \hat{k} \tag{7.3}$$

$$\hat{\imath} \cdot \hat{\imath} = \hat{\jmath} \cdot \hat{\jmath} = \hat{k} \cdot \hat{k} = 1 \tag{7.4}$$

$$\hat{\imath} \cdot \hat{\jmath} = \hat{\imath} \cdot \hat{k} = \hat{\jmath} \cdot \hat{k} = 0 \tag{7.5}$$

Using the information given in Eqs. 7.4 and 7.5, the scalar product of $\vec{A}$ and $\vec{B}$ vectors reduces to the Eq. 7.6.

$$\vec{A} \cdot \vec{B} = A_xB_x + A_yB_y + A_zB_z \tag{7.6}$$

Remembering some properties of dot product;
The result of the dot product is scalar.
The dot product of two vectors is commutative.

$$\vec{A} \cdot \vec{B} = \vec{B} \cdot \vec{A}$$

The dot product is distributive over vector addition [1, 2]

$$\vec{A} \cdot \left(\vec{B} + \vec{C}\right) = \vec{A} \cdot \vec{B} + \vec{A} \cdot \vec{C}$$

Programme 7.1 shows the scalar product of vectors $\vec{x}$ and $\vec{y}$.

The vector product, also known as the cross product, is an operation that takes two vectors in three-dimensional space and produces a third vector that is perpendicular to the plane formed by the original vectors.

The magnitude of cross product is given in Eq. 7.7.

$$\vec{A} \times \vec{B} = \left|\vec{A}\right|\left|\vec{B}\right|\sin\theta \tag{7.7}$$

Consider $\vec{A}$ and $\vec{B}$ vectors are given in forms as Eqs. 7.8 and 7.9

$$\vec{A} = A_x\hat{\imath} + A_y\hat{\jmath} + A_z \cdot \hat{k}, \tag{7.8}$$

$$\vec{B} = B_x\hat{\imath} + B_y\hat{\jmath} + B_z \cdot \hat{k} \tag{7.9}$$

```
[531]: using LinearAlgebra

[533]: x = [-2 3 ; -5 8]

[533]: 2×2 Matrix{Int64}:
        -2  3
        -5  8

[535]: y = [5 -3 ; 2 6]

[535]: 2×2 Matrix{Int64}:
        5  -3
        2   6

[543]: println("Dot (Scalar) product of x and y vectors=", dot(x, y))

       Dot (Scalar) product of x and y vectors=19
```

Programme 7.1 Scalar (dot) product of two vectors

The cross product of unit vectors $\hat{\imath}, \hat{\jmath}$ and $\hat{k}$ are represented in Eqs. 7.10 and 7.11

$$\hat{\imath} \times \hat{\imath} = \hat{\jmath} \times \hat{\jmath} = \hat{k} \times \hat{k} = 0 \tag{7.10}$$

$$\hat{\imath} \times \hat{\jmath} = \hat{k}, \hat{\jmath} \times \hat{k} = \hat{\imath}, \hat{k} \times \hat{\imath} = \hat{\jmath} \tag{7.11}$$

The cross product of $\vec{A}$ and $\vec{B}$ vectors is given by the determinant of a matrix

$$\vec{A} \times \vec{B} = \begin{vmatrix} \hat{\imath} & \hat{\jmath} & \hat{k} \\ A_x & A_y & A_z \\ B_x & B_y & B_z \end{vmatrix} \tag{7.12}$$

The result of the calculation we obtain Eq. 7.13.

$$\vec{A} \times \vec{B} = \left(A_y B_z - A_z B_y, A_z B_x - A_x B_z, A_x B_y - A_y B_x\right) \tag{7.13}$$

Remembering some properties of cross product;

The direction of resulting vector is perpendicular to $\vec{A}$ and $\vec{B}$ vectors. The direction can be determined by using the right-hand rule.

The cross product of two vectors is not commutative [1, 2].

$$\vec{A} \times \vec{B} = -\left(\vec{B} \times \vec{A}\right)$$

The cross product is distributive over vector addition.

$$\vec{A} \times \left(\vec{B} + \vec{C}\right) = \vec{A} \times \vec{B} + \vec{A} \times \vec{C}$$

The cross product is zero if the vectors are parallel or if either vector is the zero vector.

```
[19]: using LinearAlgebra

[93]: x = [2, -3, 4]

[93]: 3-element Vector{Int64}:
        2
       -3
        4

[91]: y = [-1, 5, -4]

[91]: 3-element Vector{Int64}:
       -1
        5
       -4

[95]: cross(x , y)

[95]: 3-element Vector{Int64}:
       -8
        4
        7
```

Programme 7.2 Vector product of two vectors

Programme 7.2 shows the vector product of two vectors. Programme 7.3 shows two vectors whose vector products are desired to be drawn and Programme 7.4 shows the vector products of two vectors. The ***quiver()*** function was used during the drawings. A quiver is a directed graph that consists of vertices and arrows (directed edges). Quivers are used in various fields, including representation theory, algebra. In Programmes 7.5 and 7.6, the drawing of three-dimensional vectors and their vector products are discussed [3, 4].

7.3 Falls

7.3.1 Free Fall

Free fall refers to the motion of an object falling without initial velocity under the influence of gravity alone, without any air resistance or other forces acting on it. Without air resistance or forces acting on it, all objects, regardless of their mass, will fall at the same rate due to gravity. In free fall, all objects are subject to a constant acceleration of gravity $\vec{g}$ therefore, while the acceleration of the object is constant everywhere, its velocity will increase in response to the acceleration of gravity.

Neglecting air resistance and assuming that the acceleration of free fall does not vary with height over short vertical distances, the motion of a vertically moving free-falling body is equivalent to motion in one dimension under constant acceleration.

```
[547]: using LinearAlgebra

[549]: x = [2, -3]

[549]: 2-element Vector{Int64}:
        2
       -3

[551]: y = [-1, 5]

[551]: 2-element Vector{Int64}:
       -1
        5

[555]: quiver([0,0],[0,0], quiver=([2,-1],[-3,5]))

[555]:
```

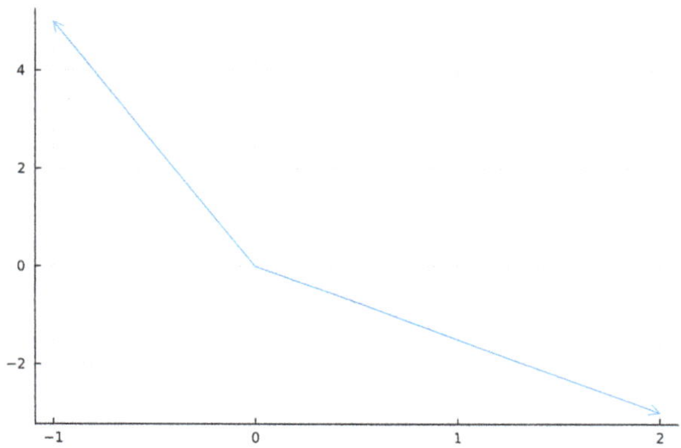

Programme 7.3 Vector representation of two vectors in 2-dimensions

```
[ ]: println(" Vector Product of x and y vectors=")

[129]: quiver([0,0],[0,0], quiver=([-1,0],[5,0]))

[129]:
```

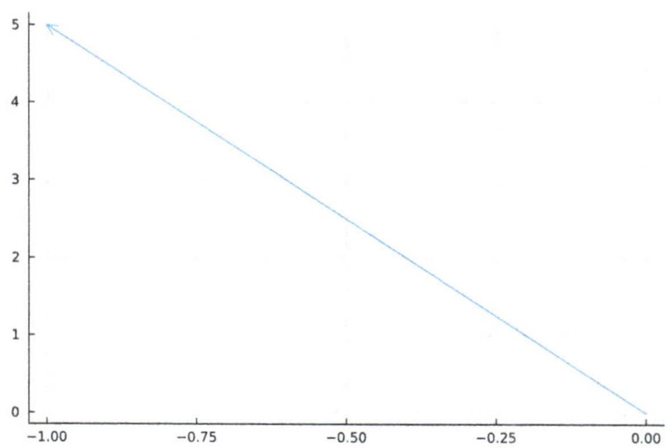

Programme 7.4 Vector representation of result vector in 2-dimensions

```
[545]: using LinearAlgebra

[131]: x = [2, -3, 4]

[131]: 3-element Vector{Int64}:
        2
       -3
        4

[133]: y = [-1, 5, -4]

[133]: 3-element Vector{Int64}:
       -1
        5
       -4

[143]: z=cross(x , y)

[143]: 3-element Vector{Int64}:
       -8
        4
        7
```

Programme 7.5 Vector product of two vectors

```
[ ]: println(" x, y  anz z vectors in three dimensions=")

[147]: quiver([0,0,0],[0,0,0],[0,0,0] , quiver=([2,-1,0-8],[-3,5,4],[-8,4,7]))

[147]:
```

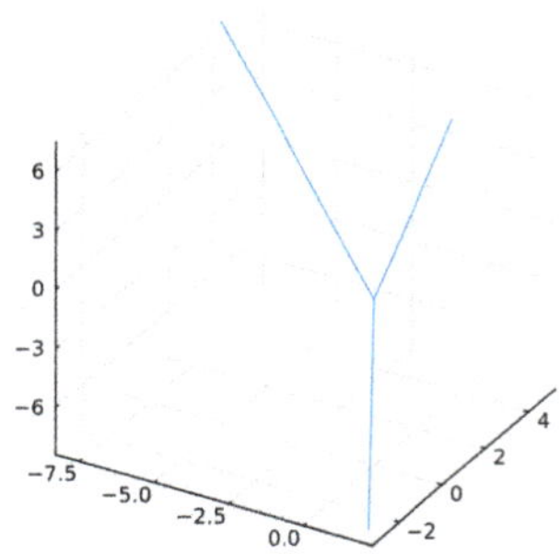

Programme 7.6 Vector product of two vectors in 3-dimensions

Therefore, the equations developed for bodies moving with constant acceleration can be applied. The only modification we need to make to these equations for free-falling bodies is to specify that the motion is in the vertical (*y-direction*) rather than the horizontal (*x-direction*) direction and that the acceleration is downward and of magnitude 9.80 m/s^2 [1, 2].

$$\overrightarrow{v_f} = \overrightarrow{v_i} + \vec{a} \cdot t \qquad v_f = v_i + a \cdot t \tag{7.14}$$

$$\overrightarrow{r_f} = \overrightarrow{r_i} + \vec{v}_i t + \frac{1}{2}\vec{a} \cdot t^2 \qquad x_f = x_i + v_i t + \frac{1}{2}a \cdot t^2 \tag{7.15}$$

In Eqs. 7.14 and 7.15, if the initial velocity $\overrightarrow{v_i}$ is taken as zero and the acceleration of the body is taken as the acceleration of gravity $\vec{g}$ and if we take the magnitude of $\overrightarrow{r_f} - \overrightarrow{r_i} = h$, the equations of free fall become as follows.

$$\overrightarrow{v_f} = \vec{g} \cdot t \qquad v_f = g \cdot t \tag{7.16}$$

$$\overrightarrow{r_f} - \overrightarrow{r_i} = \frac{1}{2}\vec{g} \cdot t^2 \qquad h = \frac{1}{2}g \cdot t^2 \tag{7.17}$$

Programme 7.7 illustrates the free fall event. First of all, it is required to define the variable data to be used in the programme for this, the command line ***val = chomp(readline))*** is used. then the initial values are assigned to the variables and the output values are given in Programme 7.8.

```
[579…  #Free Fall Example
using Plots
using SymPy
using PyCall


#determination of gravity and height
println(" input gravitation value g=")
val = chomp(readline())
g = parse(Int32, val)
#
println(" height value h=")
val = chomp(readline())
h = parse(Int32, val)

# finding time of flight
t= parse(Int32, val)
t=sqrt(2*h/g)
print("time of flight=",t,  )
#
#initial velocity
v0=0
#initial time of flight
tf=1
dis0=0
while tf <= t
  v=v0+g*tf
  dis=(1/2)g*tf^2 -(1/2)*g*(tf-1)^2
  println("velocity at time  ", tf,"second=", v, " m/s", "displacement= ",dis )

tf+=1
v0=0


end
```

Programme 7.7 Free fall example

```
 input gravitation value g=
stdin> 10
 height value h=
stdin> 80
time of flight=4.0velocity at time  1second=10 m/sdisplacement= 5.0
velocity at time  2second=20 m/sdisplacement= 15.0
velocity at time  3second=30 m/sdisplacement= 25.0
velocity at time  4second=40 m/sdisplacement= 35.0
```

Programme 7.8 Free fall output

In Programme 7.9, the variables are defined in the programme and then the equations of the free fall motion are coded. In Programmes 7.10 and 7.11, the height and velocity values are shown graphically using the ***plot()*** function, respectively [3, 4].

```
[139…  using Plots

       # Determination of variables
       g = 9.81  # Acceleration of gravity (in m/s²)
       h = 80.0 # Height (in m)
       dt = 0.5  # Time interval (in seconds)

       # Time of flight
       t_tot = sqrt(2 * h / g)

       # Interval of time
       t = 0:dt:t_tot

       # Calculation of velocity and height parameters
       y = h .- 0.5 * g .* t.^2
       v = g .* t

[139]: 0.0:4.905:39.24
```

Programme 7.9 Free fall example

```
[137]: plot(t, y, label="Height (in meters)", xlabel="Time (in seconds)", ylabel="Height (in meters)", title="Free Fall")
[137]:
```

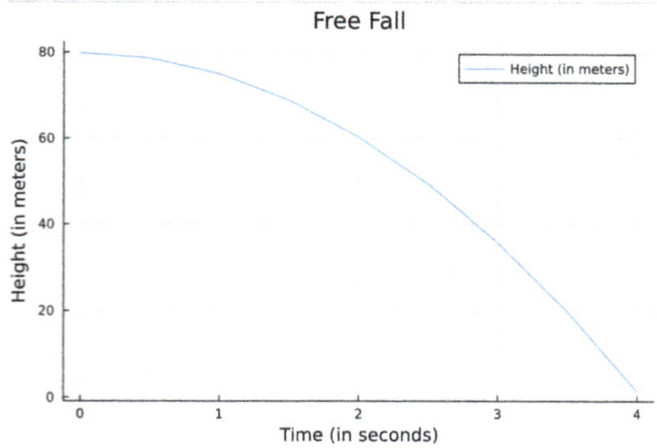

Programme 7.10 Plotting height versus time for free fall

```
[135]: plot(t, v, label="velocity(in m/s)", xlabel="Time (in seconds)", ylabel="Height (in meters)", title="Free Fall")
[135]:
```

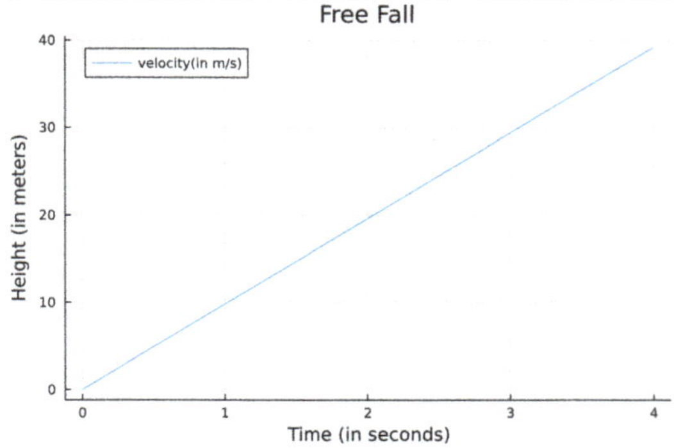

Programme 7.11 Plotting velocity versus time for free fall

7.3.2 Horizontal Motion

A horizontal motion is a physical condition in which an object is thrown in a horizontal direction and remains under the influence of gravity alone. In this type of throw, the horizontal motion of the object continues at a constant speed, while its vertical motion is accelerated due to gravity. To understand the motion of the object in a horizontal throw, we usually analyse the following two components:

Horizontal Motion: The object is assumed to move horizontally at a constant speed. This means that the object maintains the horizontal velocity it had at the beginning. The expressions and rules used for uniform linear motion will apply. Let $\boldsymbol{x}$ be the range, $\boldsymbol{v_i}$ the horizontal velocity and $\boldsymbol{t}$ the time of flight, the equation for horizontal motion

$$x = v_i \cdot t \tag{7.18}$$

can be expressed in Eq. 7.16.

Vertical Motion: The vertical motion shows an accelerated movement due to the effect of gravity. An object initially thrown from above accelerates downwards due to the effect of gravity. Therefore, free fall rules and formulas apply to the vertical motion of the object [1, 2].

$$v_f = g \cdot t \tag{7.19}$$

$$h = \frac{1}{2} g \cdot t^2 \tag{7.20}$$

$$v_f^2 = 2 \cdot g \cdot h \tag{7.21}$$

In Programme 7.12, the variables are requested to be entered as data. For this, ***val = chomp(readline())*** command is used for each variable to be entered as data.

Initial values are defined in the programme and velocity and displacement parameters are defined in the while loop and expressed in Programme 7.13 [4].

```
[9]: #horizontal  motion
     using Plots
     using SymPy
     using PyCall

     # the free fall of the problem (in vertical direction)
     #determination of gravity and height
     println(" input gravitation value g=")
     val = chomp(readline())
     g = parse(Int32, val)
     println(" input height value h=")
     val = chomp(readline())
     h = parse(Int32, val)
     println(" input horizontal velocity value v00=")
     val = chomp(readline())
     v00 = parse(Int32, val)
     # finding time of flight
     t= parse(Int32, val)
     t=sqrt(2*h/g)
     print("time of flight=",t,  )
```

Programme 7.12 Defining the variables

```
#initial horizontal velocity
v0=0
#initial time of flight
tf=0
#initial range
x=0

while tf <= t
# velocity in vertical direction
    v=v0+g*tf
    x=v00*tf
  println("velocity at time  ", tf,"second=", v, " m/s","displacement at horizontal direction",x," meters")


    tf+=1
v0=0

end
 input gravitation value g=
stdin> 10
 input height value h=
stdin> 80
 input horizontal velocity value v00=
stdin> 30
time of flight=4.0velocity at time  0second=0 m/sdisplacement at horizontal direction0 meters
velocity at time  1second=10 m/sdisplacement at horizontal direction30 meters
velocity at time  2second=20 m/sdisplacement at horizontal direction60 meters
velocity at time  3second=30 m/sdisplacement at horizontal direction90 meters
velocity at time  4second=40 m/sdisplacement at horizontal direction120 meters
```

Programme 7.13 Horizontal motion output

7.3.3 Projectile Motion

In accelerated motion in one dimension, the salt and position vectors of a body moving with constant acceleration a are treated as in Eqs. 7.22 and 7.23.

$$\vec{v_f} = \vec{v_i} + \vec{a} \cdot t \; v_{xf} = v_{xi} + a_x t \qquad v_{yf} = v_{yi} + a_y t \tag{7.22}$$

$$\vec{r_f} = \vec{r_i} + \vec{v}_i t + \tfrac{1}{2}\vec{a} \cdot t^2 \; x_f = x_i + v_{xi} t + \tfrac{1}{2} a_x \cdot t^2 \qquad y_f = y_i + v_{yi} t + \tfrac{1}{2} a_y \cdot t^2 \tag{7.23}$$

Equations 7.22 and 7.23 return when applied to the motion in two dimensions with a $\boldsymbol{v_i}$ initial velocity travelling by making $\boldsymbol{\theta}$ angle with the horizontal where θ is the angle at which the projectile leaves the origin.

The initial components of velocity are;

$$v_{xi} = v_i \cdot \cos\theta_i$$

$$v_{yi} = v_i \cdot \sin\theta_i$$

Substituting the component into Eq. 7.23 with $x_i = 0$ and $a_x = 0$, we find

$$x_f = v_{xi} \cdot t = v_i \text{cos}\theta_i \cdot t \tag{7.24}$$

If we repeat with the y component and using $y_i = 0$ and $a_y = -g$, we obtain

$$y_f = v_{yi} t + \frac{1}{2} a_y t^2 = (v_i \text{sin}\theta_i) t - \frac{1}{2} g t^2 \tag{7.25}$$

By solving the Eq. 7.24 for $t = x_f / (v_i \text{cos}\theta_i)$ and substitute this expression for t into Eq. 7.25 returns [1, 2];

$$y = (\text{tan}\theta_i) x - \left(\frac{g}{2 v_i^2 \text{cos}^2 \theta_i} \right) x^2$$

$$y = \frac{v_i^2 \text{sin}^2 \theta_i}{2g}$$

Bu using the expression $t = \frac{x_f}{v_i \text{cos}\theta_i}$, Eq. 7.24 returns;

$$x_f = \frac{v_i^2 \sin 2\theta_i}{g} \tag{7.26}$$

```
[441... #Projectile Motion
using Plots
using SymPy
using PyCall

#determination of gravity and height
println(" input gravitation value g=")
val = chomp(readline())
g = parse(Int32, val)
println(" input initial velocity v=")
val = chomp(readline())
v = parse(Int32, val)
println(" input angle alfa=")
val = chomp(readline())
alfa = parse(Int32, val)
vx=v*cos(alfa*2*pi/360)
vy=v*sin(alfa*2*pi/360)
# finding time of flight
tfu=vy/g
```

Programme 7.14 Defining the variables

```
print("time of flight=",2*tfu,"vertical initial velocity=",vy,"m/s", "horizontal initial velocity=",vx,"m/s" )
#initial velocity
h=(1/2)*g*tfu^2
print("maximum height=",h,"m" )
tf=0
x=0
while tf <= tfu
# velocity in vertical direction
    v=vy-g*tf
    x=vx*tf
    println("velocity at time", tf, "second=", v, " m/s","displacement at horizontal direction",x," meters")
    tf+=1.0
end
vy=0
n=2*(tfu)-tf
while tf > tfu && tf<=2*tfu
# velocity in vertical direction
    v=vy+g*(tf-(n+1))
    x=vx*tf
    println("velocity at time", tf, "second=", v, " m/s","displacement at horizontal direction",x," meters")
    tf+=1.0
end
```

```
 input gravitation value g=
stdin>  10
 input initial velocity v=
stdin>  50
 input angle alfa=
stdin>  37
time of flight=6.018150231520482vertical initial velocity=30.090751157602412m/shorizontal initial velocity=39.931775502364644m/smaximum heigh
t=45.272665261437545mvelocity at time0second=30.090751157602412 m/sdisplacement at horizontal direction0.0 meters
velocity at time1.0second=20.090751157602412 m/sdisplacement at horizontal direction39.931775502364644 meters
velocity at time2.0second=10.090751157602412 m/sdisplacement at horizontal direction79.86355100472929 meters
velocity at time3.0second=0.09075115760241204 m/sdisplacement at horizontal direction119.79532650709393 meters
velocity at time4.0second=9.818497684795178 m/sdisplacement at horizontal direction159.72710200945858 meters
velocity at time5.0second=19.818497684795176 m/sdisplacement at horizontal direction199.65887751182322 meters
velocity at time6.0second=29.818497684795176 m/sdisplacement at horizontal direction239.59065301418786 meters
```

Programme 7.15 Projectile motion output

In Programme 7.14, the variables are requested to be entered as data. For this, ***val = chomp(readline())*** command is used for each variable to be entered as data.

Initial values are defined in the programme and velocity and displacement parameters are defined in the while loop and expressed in Programme 7.15 [4, 5].

7.4 The Pendulum

A simple pendulum is another mechanical system that exhibits periodic motion. It consists of a particle of mass m suspended from a fixed string of length L. The motion takes place in the vertical plane and the force of gravity acts as the driving force. It is known that, provided the angle is small (less than about 10°), the motion is very close to that of a simple harmonic oscillator.

The forces acting on the mass in the system are the tensile force T applied by the string and the gravitational force mg. The tangential component of the gravitational force mgsinθ acts as a recoil force and can apply Newton's second law.

The equation of motion of mass m is given by

$$m\frac{d^2s}{dt^2} = -\text{mg } \sin\theta \tag{7.27}$$

Here ***s*** is displacement measured along the arc and also the minus sign indicates that the tangential force acts toward the equilibrium position. L is a constant and $S = L\theta$, the Eq. (7.27) reduces to

$$\frac{d^2\theta}{dt^2} = -(g/L)\sin\theta \tag{7.28}$$

However, assuming that the angle $\boldsymbol{\theta}$ is small, the equation of motion for a simple pendulum can be approximated by sinθ ~ θ;

$$\frac{d^2\theta}{dt^2} = -\frac{g}{L}\theta \tag{7.29}$$

For small oscillation amplitudes, where $\theta_{\max}$ to be the angular displacement value, the expression $\theta = \theta_{\max}\cos(wt + \varphi)$ is written for the simple harmonic motion, the Eq. 7.29 returns to

$$w = \sqrt{\frac{g}{L}} \tag{7.30}$$

where w is the angular frequency of this motion.

Therefore the period of the motion is defined by Eq. 7.31 [1, 2].

$$T = \frac{2\pi}{w} = 2\pi\sqrt{\frac{L}{g}} \tag{7.31}$$

In Programme 7.16, the variables are requested to be entered as data. For this, ***val = chomp(readline())*** command is used for each variable to be entered as data. Initial values are defined in the programme and the period of the pendulum is calculated [4, 5].

```
[1]: #the pendulum
     using SymPy
     using PyCall

     # the free fall of the problem (in vertical direction)
     #determination of gravity and height
     println(" length of pendulum (in meters)=")
     val = chomp(readline())
     l= parse(Int64, val)
     println(" acceleration of gravity (in m/s^2)=")
     val = chomp(readline())
     g= parse(Int64, val)
     pi=3
     #the period ot the motion

     T=2*pi*(sqrt(l/g))
      println("Period of the motion ", T, "in seconds")

      length of pendulum (in meters)=
     stdin> 20
      acceleration of gravity (in m/s^2)=
     stdin> 10
     Period of the motion 8.485281374238571in seconds
```

Programme 7.16 Period of the pendulum

7.5 Gravitational Acceleration

It is known that all objects are attracted towards the Earth. Gravitational force, it can be called the force exerted by the Earth on an object and this force is called the gravitational force F_g. This force is directed towards the centre of the Earth and its magnitude is called the weight.

Thus, the weight of an object defined as the magnitude of F_g is mg.

Since weight depends on $\boldsymbol{g}$ and it varies with geographical location. Therefore, unlike mass, weight is not an inherent property of an object. This is because g decreases with increasing distance from the centre of the Earth. Objects have less weight higher above sea level.

However; the acceleration of gravity within the Earth varies depending on the depth. The acceleration of gravity inside the Earth can be calculated using a simple model. The acceleration of gravity in the Earth varies with depth.

The acceleration of gravity for a given depth value d varies with *g(d)*: gravitational acceleration at depth; g_0, Gravitational acceleration at the surface (approx. 9.81 m/s^2); *R*, radius of the Earth (approximately 6.371 × 106 m). It can be formulated as Eq. 7.32.

$$g(d) = g_0\left(1 - \frac{d}{R}\right) \tag{7.32}$$

The acceleration at the surface **is** taken as $\boldsymbol{g}_0$, but as the depth decreases, the gravitational acceleration decreases. Therefore, as we approach the centre of the Earth, the gravitational acceleration approaches zero [1, 2].

```
[89]: # Gravitational acceleration (m/s^2)
      g0 = 9.81

      # Radius of the Earth (in meters)
      R = 6.371e6

      # Gravitational acceleration at depth (in meters)
      function grav_acc(d)
          if d < R
              return g0 * (1 - d / R)
          else
              return 0
          end
      end

      # Determination of the depth
      println("Enter a depth (in meters) : ")
      depth = parse(Float64, readline())

      # Ycalculation of gravitational acceleration
      g_d = grav_acc(depth)
      println("Gravitational acceleration : $g_d m/s²")

      Enter a depth (in meters) :
      stdin> 5e4
      Gravitational acceleration : 9.73301051640245 m/s²
```

Programme 7.17 Calculation of the gravitational acceleration

In Programme 7.17, variables are defined. The ***grav_acc()*** function has been defined to calculate the gravitational acceleration in the earth and the formula to be used in this function has been defined [4]. In addition, the if loop used in this function is provided to give results when values smaller than the radius of the earth are entered.

The depth value was requested from the earth and then calculations were made with the function used.

In this section, we have tried to examine the topics discussed in the mechanics section. In the programmes we have discussed, the emphasis has been on codes that are more commonly encountered and can be easily solved. When writing the programmes, we have tried to talk about different data entry and different loops. It is aimed to provide the reader with a method to work on much more pleasant problems and to show that many seemingly complex problems can be made much more understandable by using appropriate Julia codes.

References

1. Physics For Scientists and Engineers with Modern Physics, 5th Edition, Raymond A. Serway, Robert J. Beichner, Saunders College Publishing, 2000.; Gravitational acceleration, Chapter 14; the pendulum, Chapter 13; Falls. Chapter 2 4, Scalar and Vector Product of Vectors chapter 3
2. Foundamentals of Physics (Extended Edition), David Halliday and Robert Reisnick, 10th Edition. ISBN 978-1-11823072-5. 2007.; Gravitational acceleration, Chapter 13; the pendulum, Chapter 15; Falls. Chapter 4, Scalar and Vector Product of Vectors chapter 3
3. Julia-Plotting Vector Fields, https://discourse.julialang.org/t/plotting-vector-fields/32704
4. Julia-Manual-Functions. https://docs.julialang.org/en/v1/manual/functions/
5. Julia-Develloper Documantations- Documentations of Julia's Internals- Julia Functions. https://docs.julialang.org/en/v1/devdocs/functions/

Chapter 8
Julia Programming Applications in Electromagnetism

8.1 Introduction

In this section, we will focus on the basic problems from simple to difficult from a pedagogical point of view related to the subject called waves in physics and the analysis of these problems using the Julia Programming language. After some applications, more detailed parts of the analysis will be left as questions for the reader. Thus, it is thought that both the Julia programming language and, more importantly, the physics problem will be discussed in a more understandable way.

8.2 Electrical Field of a Charge

An electric field is a region around a charged particle where other charged particles experience a force. It is a vector field, meaning it has both magnitude and direction. The electric field at a point in space is defined as the electric force acting on a positive test charge placed at that point divided by the magnitude of the test charge. The electric field $\vec{E}$ created by a charge can be described mathematically as:

$$\vec{E} = \frac{\vec{F}}{q} \tag{8.1}$$

where $\vec{F}$ is the force experienced by a small positive test charge q.

The direction of the electric field is defined as the direction of the force that a positive test charge would experience. The electric field vector is in the direction of the force acting on a positive test charge placed in the field. Assuming that the electric force acts on a positive test charge, the electric field will be in the direction of the electric force. It points away from positive charges and toward negative charges.

R. G. Türeci et al., *Julia Programming for Physics Applications*,
https://doi.org/10.1007/978-3-031-84716-5_8

The strength of the electric field decreases with the square of the distance from the charge, according to Coulomb's law. For a point charge Q, the electric field at a distance $\vec{r}$ is given by:

$$\vec{F} = k \cdot \frac{Q \cdot q}{r^2} \quad (8.2)$$

$$\left|\vec{E}\right| = k \cdot Q/r^2 \quad (8.3)$$

The SI unit of electric field strength is volts per meter (V/m) [1, 2].

In Programme 8.1, first we define the variables used in Programme 8.1, the ***parse*** function is used to convert a string representation of a value into the specified data type. This is especially useful for converting strings to numbers, dates, or other types.

You can create a mesh grid using the ***meshgrid*** function from the LinearAlgebra standard library. A mesh grid is useful for evaluating functions over a grid of values, commonly used in plotting and numerical computations.

The ***collect()*** function is used to convert an iterable (like a range, or any other collection) into an actual array or collection type. This is particularly useful when you want to create an array from a sequence of values or to force the evaluation of an iterable.

Programme 8.2 shows the output values for negative charge.

```
[197]: using Plots

[243]: # Constants
       k = 8.99e9  # Coulomb's constant (N m²/C²)
       charge_position = (0, 0)  # Position of the charge
       println("Enter charge (in e-x form) (in Coulombs): ")
       q = parse(Float64, readline())
       # Create a grid of points
       x = -10:0.5:10
       y = -10:0.5:10

       # Labels and title of the graph

       xlims!(-10, 10)
       ylims!(-10, 10)

       # Create mesh grid
       X = collect(x)
       Y = collect(y)

       # Calculate the electric field vectors
       X_grid = X .+ zeros(length(y), length(x))
       Y_grid = zeros(length(y), length(x)) .+ Y
```

Programme 8.1 Defining the variables and creating mesh grid

```
        Enter charge (in e-x form) (in Coulombs):
        stdin>  -5e-10
[243]:  41×41 Matrix{Float64}:
         -10.0  -10.0  -10.0  -10.0  -10.0  …  -10.0  -10.0  -10.0  -10.0  -10.0
          -9.5   -9.5   -9.5   -9.5   -9.5      -9.5   -9.5   -9.5   -9.5   -9.5
          -9.0   -9.0   -9.0   -9.0   -9.0      -9.0   -9.0   -9.0   -9.0   -9.0
          -8.5   -8.5   -8.5   -8.5   -8.5      -8.5   -8.5   -8.5   -8.5   -8.5
          -8.0   -8.0   -8.0   -8.0   -8.0      -8.0   -8.0   -8.0   -8.0   -8.0
          -7.5   -7.5   -7.5   -7.5   -7.5  …   -7.5   -7.5   -7.5   -7.5   -7.5
          -7.0   -7.0   -7.0   -7.0   -7.0      -7.0   -7.0   -7.0   -7.0   -7.0
          -6.5   -6.5   -6.5   -6.5   -6.5      -6.5   -6.5   -6.5   -6.5   -6.5
          -6.0   -6.0   -6.0   -6.0   -6.0      -6.0   -6.0   -6.0   -6.0   -6.0
          -5.5   -5.5   -5.5   -5.5   -5.5      -5.5   -5.5   -5.5   -5.5   -5.5
          -5.0   -5.0   -5.0   -5.0   -5.0  …   -5.0   -5.0   -5.0   -5.0   -5.0
          -4.5   -4.5   -4.5   -4.5   -4.5      -4.5   -4.5   -4.5   -4.5   -4.5
          -4.0   -4.0   -4.0   -4.0   -4.0      -4.0   -4.0   -4.0   -4.0   -4.0
           ⋮                                ⋱                                ⋮
           4.5    4.5    4.5    4.5    4.5       4.5    4.5    4.5    4.5    4.5
           5.0    5.0    5.0    5.0    5.0  …    5.0    5.0    5.0    5.0    5.0
           5.5    5.5    5.5    5.5    5.5       5.5    5.5    5.5    5.5    5.5
           6.0    6.0    6.0    6.0    6.0       6.0    6.0    6.0    6.0    6.0
           6.5    6.5    6.5    6.5    6.5       6.5    6.5    6.5    6.5    6.5
           7.0    7.0    7.0    7.0    7.0       7.0    7.0    7.0    7.0    7.0
           7.5    7.5    7.5    7.5    7.5  …    7.5    7.5    7.5    7.5    7.5
           8.0    8.0    8.0    8.0    8.0       8.0    8.0    8.0    8.0    8.0
           8.5    8.5    8.5    8.5    8.5       8.5    8.5    8.5    8.5    8.5
           9.0    9.0    9.0    9.0    9.0       9.0    9.0    9.0    9.0    9.0
           9.5    9.5    9.5    9.5    9.5       9.5    9.5    9.5    9.5    9.5
          10.0   10.0   10.0   10.0   10.0  …   10.0   10.0   10.0   10.0   10.0
```

Programme 8.2 Output for negative charge

In Programme 8.3, ***quiver*** plot is used to visualise vector fields, showing direction and magnitude of vectors at specified points in a 2D or 3D space. You can create quiver plots using various plotting libraries, such as ***Plots.jl*** or ***Makie.jl***.

```
[245]:
R = sqrt.(X_grid.^2 .+ Y_grid.^2)  # Distance from the charge
E_mag = k * q ./ (R .^ 2)  # Electric field magnitude

E_mag[R .== 0] .= 0

# Calculate the electric field components
Ex = E_mag .* (X_grid ./ R)  # X-component
Ey = E_mag .* (Y_grid ./ R)  # Y-component

# Create a quiver plot to visualize the electric field
quiver(X_grid, Y_grid, quiver=(Ex,Ey),  color=:blue, label="Electric Field")

# Charge representation
scatter!([charge_position[1]], [charge_position[2]], marker=:star, color=:red, label="Charge")

xlabel!("X (in meters)")
ylabel!("Y (in meters)")
title!("Electric Field of a Point Charge")
```

Programme 8.3 Calculation of electric field

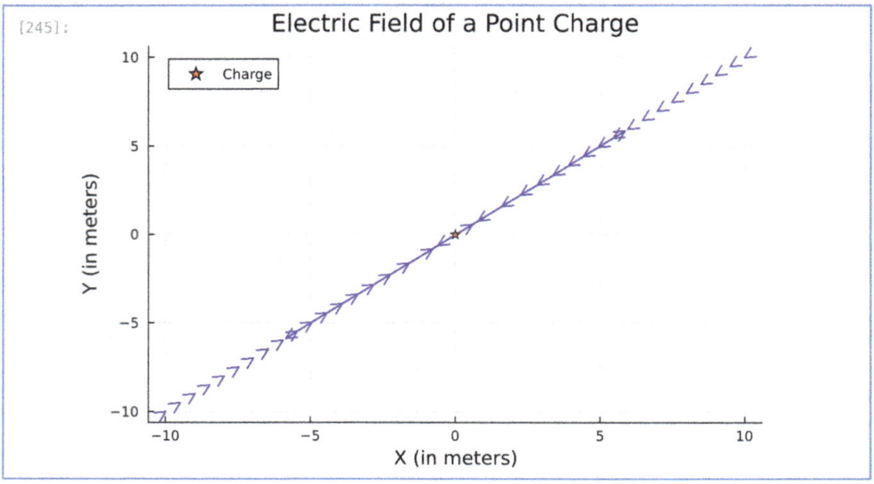

Programme 8.4 The representation of electric field for negative charge

A ***scatter*** plot is used to visualise the relationship between two or more variables by displaying individual data points on a Cartesian coordinate system. You can create scatter plots using popular plotting libraries such as ***Plots.jl*** or ***Makie.jl***.

In Programme 8.4, symbolise*s* vector fields, showing direction and magnitude of vectors at specified points in a 2D for a negative charge [3].

Programmes 8.5 and 8.6 symbolise the data for a positive charge and electric field diagram for the positive charge [3].

Programme 8.4 symbolises the electric field vector for a negative charge.

Programme 8.5 shows the output values for positive charge.

Programme 8.6 symbolises the electric field vector for a positive charge.

8.3 Lorentz Force

A charge moving with velocity $\vec{v}$ in both an electric field $\vec{E}$ and a magnetic field $\vec{B}$ moves under the influence of the electric force $q \cdot \vec{E}$ and the magnetic force $q \cdot \vec{v} \times \vec{B}$. The total force acting on the charge is the Lorentz force.

$$\vec{F} = q \cdot \vec{E} + q \cdot \left(\vec{v} \times \vec{B}\right) \tag{8.4}$$

Let us consider a uniform electric field directed vertically downwards and a uniform magnetic field perpendicular to the electric field in the plane of the page. Assuming that the charge ***q*** is positive, a downward electric force and an upward magnetic force will act on this charge during its motion. If the magnitudes of the two fields are chosen to be $q \cdot E = q \cdot v \cdot B$ such that the particle travels along a

```
         Enter charge (in e-x form) (in Coulombs):
         stdin>  2e-6
[311]:   41×41 Matrix{Float64}:
          -10.0  -10.0  -10.0  -10.0  -10.0  …  -10.0  -10.0  -10.0  -10.0  -10.0
           -9.5   -9.5   -9.5   -9.5   -9.5      -9.5   -9.5   -9.5   -9.5   -9.5
           -9.0   -9.0   -9.0   -9.0   -9.0      -9.0   -9.0   -9.0   -9.0   -9.0
           -8.5   -8.5   -8.5   -8.5   -8.5      -8.5   -8.5   -8.5   -8.5   -8.5
           -8.0   -8.0   -8.0   -8.0   -8.0      -8.0   -8.0   -8.0   -8.0   -8.0
           -7.5   -7.5   -7.5   -7.5   -7.5  …   -7.5   -7.5   -7.5   -7.5   -7.5
           -7.0   -7.0   -7.0   -7.0   -7.0      -7.0   -7.0   -7.0   -7.0   -7.0
           -6.5   -6.5   -6.5   -6.5   -6.5      -6.5   -6.5   -6.5   -6.5   -6.5
           -6.0   -6.0   -6.0   -6.0   -6.0      -6.0   -6.0   -6.0   -6.0   -6.0
           -5.5   -5.5   -5.5   -5.5   -5.5      -5.5   -5.5   -5.5   -5.5   -5.5
           -5.0   -5.0   -5.0   -5.0   -5.0  …   -5.0   -5.0   -5.0   -5.0   -5.0
           -4.5   -4.5   -4.5   -4.5   -4.5      -4.5   -4.5   -4.5   -4.5   -4.5
           -4.0   -4.0   -4.0   -4.0   -4.0      -4.0   -4.0   -4.0   -4.0   -4.0
             ⋮                                ⋱                               ⋮
            4.5    4.5    4.5    4.5    4.5       4.5    4.5    4.5    4.5    4.5
            5.0    5.0    5.0    5.0    5.0  …    5.0    5.0    5.0    5.0    5.0
            5.5    5.5    5.5    5.5    5.5       5.5    5.5    5.5    5.5    5.5
            6.0    6.0    6.0    6.0    6.0       6.0    6.0    6.0    6.0    6.0
            6.5    6.5    6.5    6.5    6.5       6.5    6.5    6.5    6.5    6.5
            7.0    7.0    7.0    7.0    7.0       7.0    7.0    7.0    7.0    7.0
            7.5    7.5    7.5    7.5    7.5  …    7.5    7.5    7.5    7.5    7.5
            8.0    8.0    8.0    8.0    8.0       8.0    8.0    8.0    8.0    8.0
            8.5    8.5    8.5    8.5    8.5       8.5    8.5    8.5    8.5    8.5
            9.0    9.0    9.0    9.0    9.0       9.0    9.0    9.0    9.0    9.0
            9.5    9.5    9.5    9.5    9.5       9.5    9.5    9.5    9.5    9.5
           10.0   10.0   10.0   10.0   10.0  …   10.0   10.0   10.0   10.0   10.0
```

Programme 8.5 Output for positive charge

[319]:

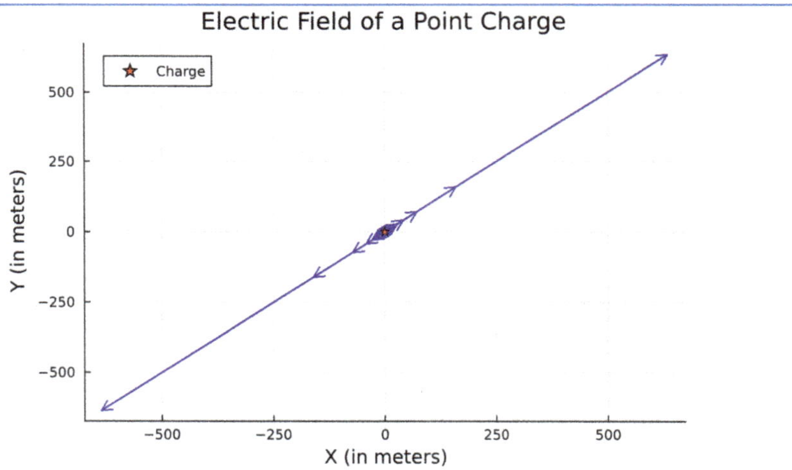

Programme 8.6 The representation of electric field for the positive charge

linear horizontal line in the region where the fields are located. From the expression $q \cdot E = q \cdot v \cdot B$,

$$v = \frac{E}{B} \tag{8.5}$$

is obtained.

Only particles with this speed can pass through electric and magnetic fields perpendicular to each other without deviation.

The Lorentz force is the basis of mass spectrometry. The mass spectrometer is a device that separates masses according to the ratio of their charges. Charged particles are first passed through a speed selector and then enter a magnetic field. Thanks to the magnetic force acting on the particle in the magnetic field, the particle draws a semi-circular trajectory and hits the photographic film [1, 2] (Programmes 8.7 and 8.8).

```
[ ]: using Plots
     using LinearAlgebra

[249]:
     # Constants
     println("Enter charge (in e-x form) (in Coulombs): ")
     q = parse(Float64, readline())
     # Electric field (N/C) definition
     E = [1.0, 0.0, 1.0]
     # Magnetic field (T) definition
     B = [0.0, 1.0, 0.0]
     # Initial velocity (in m/s)
     v0 = [1.0, 0.0, 0.0]

     # Time parameters
     t_max = 2  # Maximum time (s)
     dt = 0.2    # Time step (s)
     t = 0:dt:t_max

     Enter charge (in e-x form) (in Coulombs):
     stdin> -3e-6
[249]: 0.0:0.2:2.0
```

Programme 8.7 Definition of variables for Lorentz force

```
[323]: positions = zeros(length(t), 3)  # Store x, y, z positions

       v = copy(v0)  # Current velocity

       # Calculate the position of the particle over time
       for i in 1:length(t)
           # Lorentz force
           F = q * (E + cross(v, B))

           # a = F/m
           a = F / q
           v += a * dt

           if i > 1
               positions[i, :] = positions[i-1, :] + v * dt
           end
       end
```

Programme 8.8 Programme part of the Lorentz force

```
[321]: # Plot the trajectory
       plot3d(positions[:, 1], positions[:, 2], positions[:, 3], label="Trajectory",
           xlabel="X (in meters)", ylabel="Y (in meters)", zlabel="Z (in meters)", title="Effect of Lorentz Force")

       # Normalize the tangent vectors for uniform arrow length
       tangent_norms = sqrt.(sum(tangents.^2, dims=2))
       normalized_tangents = tangents ./ tangent_norms

       #  Arrow positions
       arrow_positions = positions[1:end-1, :]
       arrow_ends = arrow_positions .+ 0.2 .* normalized_tangents

       # Add direction arrows
       quiver!(arrow_positions[:, 1], arrow_positions[:, 2], arrow_positions[:, 3],
               quiver=(arrow_ends[:, 1] - arrow_positions[:, 1],
                       arrow_ends[:, 2] - arrow_positions[:, 2],
                       arrow_ends[:, 3] - arrow_positions[:, 3]),
               color=:darkred, label="Direction Arrows", length=0.2, arrowhead=10, lw=2)

       # Show the plot
       display(current())
```

Programme 8.9 Plotting the trajectory

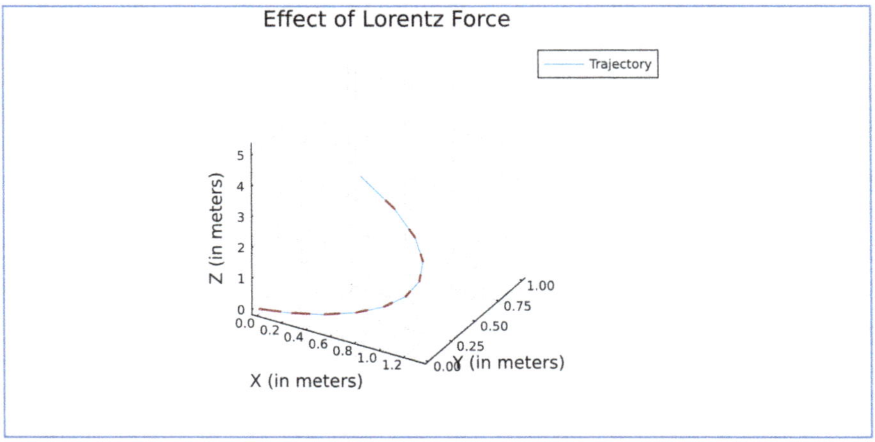

Programme 8.10 Trajectory of a particle

Programmes 8.9 and 8.10 symbolises ***plot3d()*** functions and trajectory [3–5].

8.4 Series and Parallels Circuits

If two or more resistors are connected to each other in such a way that they have only one common point, these resistors are called series connected resistors. In series connections, all loads moving through one resistor must also pass through the second resistor. Let's consider the circuit in which resistors R_1 and R_2 are present and connected to each other in series. Since the load flowing through resistor R_1 must be equal to the load flowing through resistor R_2 in series connection of resistors,

```
[327]: using LinearAlgebra

       # Define series circuits parameters
       println("Enter potentiel difference(in Volts): ")
       V_i = parse(Float64, readline())
       println("Enter resistor 1 value (in Ohm): ")
       R1 = parse(Float64, readline())
       println("Enter resistor 2 value (in Ohm): ")
       R2 = parse(Float64, readline())

       # I_total = Current through R1 and R2

       # Rearranging gives us:
       # I_total = V_i / (R1 + R2)

       I_total = V_i / (R1+ R2)

       # Print results
       println("Total Current (I_total): ", I_total, " A")

       # Calculate voltage across each resistor using KVL
       V_R1 = I_total * R1
       V_R2 = I_total * R2

       println("Voltage across R1: ", V_R1, " V")
       println("Voltage across R2: ", V_R2, " V")

       Enter potentiel difference(in Volts):
       stdin> 20
       Enter resistor 1 value (in Ohm):
       stdin> 4
       Enter resistor 2 value (in Ohm):
       stdin> 7
       Total Current (I_total): 1.8181818181818181 A
       Voltage across R1: 7.2727272727272725 V
       Voltage across R2: 12.727272727272727 V
```

Programme 8.11 Serial circuit example

the current flowing through both resistors will be the same. However, the potential difference applied between the ends of the series connected resistors will be divided between the resistors. The potential drop between the ends of resistor R_1 is equal to $I \cdot R_1$ and the potential drop between the ends of resistor R_2 is equal to $I \cdot R_2$.

The potential difference between the ends of all series connected resistors can be expressed as

$$\Delta V = IR_1 + IR_2 \tag{8.6}$$

Therefore, we can replace all series resistors with one equivalent resistor.

$$R_{\text{eq}} = R_1 + R_2 \tag{8.7}$$

This means that when the R_{eq} resistor is used instead of the $R_1 + R_2$ resistor, the current in the circuit will remain unchanged.

Now let's examine parallel connected resistors. If the main circuit current divides into two or more branches at a point, this point is called a node. In this case, the current passing through resistor R_1 is I_1, and the current passing through resistor R_2 is I_2. The potential differences of the resistors on parallel branches will be the same. Because the resistors connected in parallel are connected directly to the terminals of the battery. The smaller current will flow through the larger resistor. Therefore, the current I in the main branch will be divided when it reaches the node. Since the load

```
[329]: using LinearAlgebra

       # Define circuit parameters
       println("Enter potentiel difference(in Volts): ")
       V_i = parse(Float64, readline())
       println("Enter resistor 1 value (in Ohm): ")
       R1 = parse(Float64, readline())
       println("Enter resistor 2 value (in Ohm): ")
       R2 = parse(Float64, readline())

       # Calculate the current through each resistor
       I1 = V_i / R1   # Current through R1
       I2 = V_i/ R2   # Current through R2

       # Total current
       I_total = I1 + I2

       # Print results
       println("Current through R1 (I1): ", I1, " A")
       println("Current through R2 (I2): ", I2, " A")
       println("Total Current (I_total): ", I_total, " A")

       # Verify voltage across each resistor
       V_R1 = I1 * R1
       V_R2 = I2 * R2

       println("Voltage across R1: ", V_R1, " V")
       println("Voltage across R2: ", V_R2, " V")

       Enter potentiel difference(in Volts):
       stdin> 20
       Enter resistor 1 value (in Ohm):
       stdin> 5
       Enter resistor 2 value (in Ohm):
       stdin> 4
       Current through R1 (I1): 4.0 A
       Current through R2 (I2): 5.0 A
       Total Current (I_total): 9.0 A
       Voltage across R1: 20.0 V
       Voltage across R2: 20.0 V
```

Programme 8.12 Parallel circuit example

must be protected, the sum of the currents arriving at the node must be equal to the sum of the currents divided at the node.

$$I = I_1 + I_2 \tag{8.8}$$

Since the potential difference between the ends of each resistor will be the same as stated in Eq. 8.6,

$$\begin{aligned} I &= I_1 + I_2 \\ &= \frac{\Delta V}{R_1} + \frac{\Delta V}{R_2} = \Delta V\left(\frac{1}{R_1} + \frac{1}{R_2}\right) = \frac{\Delta V}{R_{\text{eq}}} \end{aligned} \tag{8.9}$$

returns to Eq. 8.9 (Programmes 8.11, 8.12 and 8.13).

From here, the equivalent resistance of two resistors connected in parallel can be given in Eq. 8.10 [1, 2]

```
using Plots

const k = 8.99e9  # N m²/C²
# Define a structure for charge
struct Charge
    position::Vector{Float64}
    value::Float64
end

# Function to calculate the force between two charges
function calculate_force(q1::Charge, q2::Charge)
    r_vec = q2.position - q1.position
    r_mag = norm(r_vec)
    if r_mag == 0
        return [0.0, 0.0]
    end
    # Calculate the force magnitude using Coulomb's Law
    f_mag = k * abs(q1.value * q2.value) / r_mag^2
    f_dir = r_vec / r_mag  # Direction of the force
    return f_mag * f_dir
end

# Calculate the net force
function net_force(charges::Vector{Charge})
    forces = zeros(Float64, length(charges), 2)
    for i in 1:length(charges)
        for j in 1:length(charges)
            if i != j
                forces[i, :] += calculate_force(charges[i], charges[j])
            end
        end
    end
    return forces
end
```

Programme 8.13 Defining the variables and functions

$$\frac{1}{R_{eq}} = \left(\frac{1}{R_1} + \frac{1}{R_2}\right) \tag{8.10}$$

8.5 Electrical Force Between Two Charges

A quantitative measurement of the electrical repulsion or attraction forces between charged objects was provided by Charles Coulomb, using a torsion pendulum by replacing masses with charged spheres.

Coulomb experiments show that the electrical force between two stationary charged particles;

is oriented along the line connecting the particles and is inversely proportional to the square of the distance between them, directly proportional to the product of the charges $\boldsymbol{q_1}$ and $\boldsymbol{q_2}$ on the spheres and that the force is attractive for charges with opposite signs and repulsive for charges with the same sign.

Based on these observations, Coulomb's law; We can express the magnitude of the electrical force between two point charges as in Eq. 8.11.

$$F = k\frac{|q_1||q_2|}{r^2} \tag{8.11}$$

```
# Define some charges
charges = [
    Charge([1.0, 3.0], +2.0),     # Charge  (2 C)
    Charge([3.0, 5.0], +3.0),    # Charge  (+3 C)

]

# Calculate net forces
forces = net_force(charges)

# Print results
for i in 1:length(charges)
    println("Net Force ", i, ": ", forces[i, :], " N")
end

# Plot the charges and the force vectors
scatter([c.position[1] for c in charges], [c.position[2] for c in charges],
        label="Charges", legend=:topright, xlabel="X (m)", ylabel="Y (m)",
        title="Electrical Forces ", marker=:star, markersize=2)

# Add force vectors to the plot
for i in 1:length(charges)
    arrow_start = charges[i].position
    arrow_end = charges[i].position .+ forces[i, :] * 1e-10
    plot!([arrow_start[1], arrow_end[1]], [arrow_start[2], arrow_end[2]],
          label="", color=:blue, lw=2, arrow=true)
end

# Show the plot
display(current())

Net Force 1: [4.767667472150295e9, 4.767667472150295e9] N
Net Force 2: [-4.767667472150295e9, -4.767667472150295e9] N
```

Programme 8.14 Defining the charges and calculating the forces

In the SI unit system, k is the Coulomb constant and $\in_0$ is the electrical permittivity of empty space, can be defined in Eq. 8.12.

$$k = 8.9875 \times 10^9 \text{ N m}^2/\text{C}^2 \tag{8.12}$$

In operations related to Coulomb's law [1, 2], it should not be forgotten that force is a vector quantity. If we write the expression as a vector, $\hat{r}$ in the equation is a vector directed from the charge q_1 to the charge q_2. Since the electric force obeys Newton's third law, the force applied by q_1 to q_2 and the force applied by q_2 to q_1 are equal in magnitude but opposite in direction.

$$\vec{F} = k\frac{q_1 q_2}{r^2}\hat{r} \tag{8.13}$$

$$\vec{F}_{12} = -\vec{F}_{21} \tag{8.14}$$

Programmes 8.14 and 8.15 symbolise *plot()* function and its variables [3–5].

In this section, we dealt with the problems we will encounter in electromagnetism. We made use of charges without the presence of electrical forces and electric fields. By using magnets instead of loads, the same processes can be repeated and programs related to magnetic fields and magnetic forces can be written using formulas related to the subject. We tried to touch upon both calculations and chart drawing in this section.

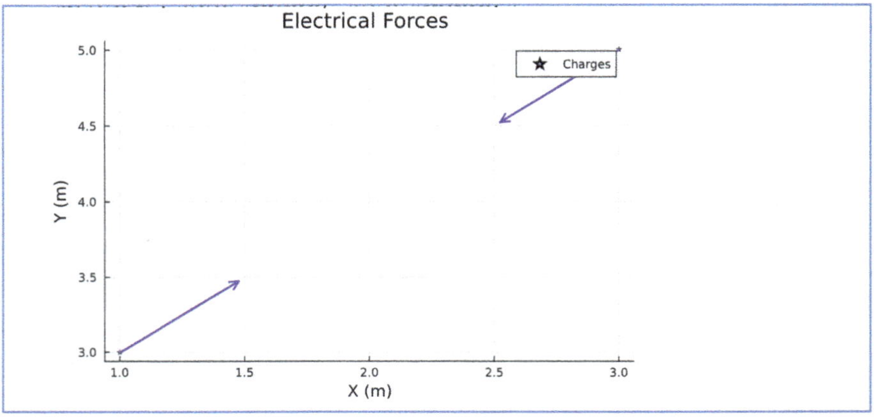

Programme 8.15 Demonstration of electrical force

References

1. Physics For Scientists and Engineers with Modern Physics, 5th Edition, Raymond A. Serway, Robert J. Beichner, Saunders College Publishing, 2000.; Electrical Field of a Charge, Chapter 23, Lorentz Force, Chapter 31, Electrical Force Between Two Charges, Chapter 22, Series and Parallels Curcits, Chapter 28.
2. Foundamentals of Physics (Extended Edition), David Halliday and Robert Reisnick, 10th Edition. ISBN 978-1-11823072-5. 2007.; Electrical Field of a Charge, Chapter 22, Lorentz Force, Chapter 22, 28 Electrical Force Between Two Charges, Chapter 21, Series and Parallels Curcits, Chapter 26, 27
3. Julia- Plotting Vector Fields. https://discourse.julialang.org/t/plotting-vector-fields/32704
4. Julia- Manual- Functions, https://docs.julialang.org/en/v1/manual/functions/
5. Julia- Developer Documentation- Documentatin of Julia's Internals- Julia Functions, https://docs.julialang.org/en/v1/devdocs/functions/

Chapter 9
Julia Programming Applications in Waves

9.1 Introduction

In this section, we will focus on the basic problems from simple to difficult from a pedagogical point of view related to the subject called waves in physics and the analysis of these problems using the Julia programming language. After some applications, more detailed parts of the analysis will be left as questions for the reader. Thus, it is thought that both the Julia programming language and, more importantly, the physics problem will be discussed in a more understandable way. The reader is referred to the elegant work of Thornton and Marion [1] for a detailed review of the issues to be discussed in this chapter.

9.2 Mass-Spring System in the Absence of Friction Effect

Suppose that a body of mass $\boldsymbol{m}$ attached to the end of a spring with spring constant $\boldsymbol{k}$ on horizontal ground is at rest in the equilibrium position. Taking the equilibrium position of the mass as the origin, we want to study the motion of this mass when the mass is pulled from the equilibrium position by $\boldsymbol{x}$ and released. According to Newton's second law and Hooke's law, we need to write and solve the equation of motion of the mass at the end of the spring. The solution of the equation of motion will give us the change of the position of the mass with respect to time. Here we also want to calculate the kinetic energy, potential energy and total energy of the mass and even graph the position, kinetic energy, potential energy and total energy with respect to time. The equation of motion of mass m is given by

$$m\frac{d^2x(t)}{dt^2} = -kx(t). \tag{9.1}$$

R. G. Türeci et al., *Julia Programming for Physics Applications*,
https://doi.org/10.1007/978-3-031-84716-5_9

Here ***k*** is a positive number corresponding to the spring constant, ***m*** to the mass, ***x*** to the position and ***t*** to the time variable. The solution of Eq. (9.1) will give us ***x(t)***. For the mass-spring system, the energy stored in the spring and the kinetic energy of the mass are given by

$$U = \frac{1}{2}kx(t)^2, \tag{9.2}$$

$$T = \frac{1}{2}m\left(\frac{dx(t)}{dt}\right)^2 = \frac{1}{2}m\dot{x}(t)^2, \tag{9.3}$$

respectively, where $\dot{x}(t)$ means the first order time derivative of $\boldsymbol{x}$.

Now let's try to solve the problem we defined above with the Julia programming language [2]. In fact, our problem here is an initial value problem. We assume that the mass is pulled 4 cm from its equilibrium position and then released. Therefore $x(t = 0) = 4$ and $\left.\frac{dx(t)}{dt}\right|_{t=0} = \dot{x}(t = 0) = 0$.

Let's start by defining our variables and function with the *SymPy* module [3] previously discussed in Chap. 5, Programme 9.1. With *SymPy* in Julia language, the differential equation given in Programme 9.1 can be written as in Programme 9.2. and the solution of this differential equation is obtained with the ***dsolve()*** function in the same program.

After naming the right-hand side of the solution of the differential equation as the function ***f***, we first use ***free_symbols(f)*** [1] to calculate the initial value of how far the mass leaves the equilibrium position for ***t = 0***. Then we substitute the obtained result into the ***f*** function. Since there is only one integral constant left in the equation, we obtain the solution to the problem we defined above by using the second initial

```
[2]: using SymPy

[3]: t, ω0 = symbols(:t), symbols(:ω0)

[3]: (t, ω0)

[4]: x = SymFunction("x")(t)

[4]: x(t)
```

Programme 9.1 Defining the variables and the function

```
[5]: diff_eq = Eq(diff(x(t), t, t), - ω0^2 * x)

[5]: d²/dt² x(t) = −ω0² x(t)

[6]: sol = dsolve(diff_eq, x(t))

[6]: x(t) = C1 e^(−itω0) + C2 e^(itω0)
```

Programme 9.2 The differential equation given by Eq. 9.1 and its solution

```
[7]: f = sol.rhs()
```

[7]: $C_1e^{-it\omega_0} + C_2e^{it\omega_0}$

```
[8]: c1 = free_symbols(f)[1]
```

[8]: C_1

```
[9]: c1v = solve(f(t => 0) ~ 4, c1)[1]
```

[9]: $4 - C_2$

```
[10]: f = f(c1 => Sym(c1v[1]))
```

[10]: $C_2e^{it\omega_0} + (4 - C_2)e^{-it\omega_0}$

```
[11]: f
```

[11]: $C_2e^{it\omega_0} + (4 - C_2)e^{-it\omega_0}$

```
[12]: C2 = free_symbols(f)[1]
```

[12]: C_2

```
[13]: C2v = solve(diff(f, t)(t => 0), free_symbols(f)[1])
```

[13]: [2]

```
[14]: solv = f(C2 => Sym(C2v[1]))
```

[14]: $2e^{it\omega_0} + 2e^{-it\omega_0}$

```
[15]: X = SymFunction("X")(t)
      X = simplify(solv)
      X
```

[15]: $4\cos(t\omega_0)$

```
[16]: X(t)
```

[16]: $4\cos(t\omega_0)$

Programme 9.3 The solution of the differential equation and applying the initial values

value with ***free_symbols(f)*** [1] and simplifying the result with the ***simplify()*** function (Programme 9.3).

Now, by defining potential energy and kinetic energy as a function of time, we can examine the variation of potential energy and kinetic energy with time by substituting ***k/m*** for angular frequency. Since we ignore friction, it is clear that the total energy will remain constant, but at any instant ***t*** the sum of kinetic energy and potential energy will give the total energy. These calculations are given in Programme 9.4 (Programme 9.5).

9.3 Damped Oscillation

We now consider the mass-spring system studied in the previous section, but this time we take into account the presence of a friction that resists the motion of the mass. Here we consider the effect of friction as a term proportional to the velocity. We again assume that the mass is separated from the equilibrium position by 4 cm

```
[17]:  m, k = symbols(:m), symbols(:k)
       U = SymFunction("U")(t)
       T = SymFunction("T")(t)

       X = subs(X, (w0, sqrt(k/m)), Dict((k => 2), (m => 1)))
```

[17]: $4\cos\left(\sqrt{2}t\right)$

```
[18]:  U = 1/2 * k * X^2
```

[18]: $8.0k\cos^2\left(\sqrt{2}t\right)$

```
[19]:  U = subs(U, (w0, sqrt(k/m)), Dict((k => 2), (m => 1)))
```

[19]: $16.0\cos^2\left(\sqrt{2}t\right)$

```
[20]:  T = 1/2 * m * diff(X(t), t)^2
```

[20]: $16.0m\sin^2\left(\sqrt{2}t\right)$

```
[21]:  T = subs(T, (w0, sqrt(k/m)), Dict((k => 2), (m => 1)))
```

[21]: $16.0\sin^2\left(\sqrt{2}t\right)$

```
[22]:  En = T + U
```

[22]: $16.0\sin^2\left(\sqrt{2}t\right) + 16.0\cos^2\left(\sqrt{2}t\right)$

```
[23]:  simplify(subs(En, (w0^2, k/m)))
```

[23]: 16.0

Programme 9.4 The potential energy, kinetic energy and the sum of these energies

and that at $\boldsymbol{t = 0}$ the mass is released. Equation of motion of the system is given by

$$\frac{d^2x(t)}{dt^2} = -2b\frac{dx(t)}{dt} - \frac{k}{m}x(t). \tag{9.4}$$

We want to solve the differential equation given by Eq. (9.4) using the *SymPy* module in Julia language. Here $b = k\big/(2m)$ is the damping parameter and is included in the solution. Three different solutions are obtained for the possible cases of the damping parameter and angular frequency. These solutions are critical damping with $\omega_0^2 = b^2$, over damping with $\omega_0^2 < b^2$ and subcritical damping with $\omega_0^2 > b^2$. In Programme 9.6, the *SymPy* module was loaded and the differential equation given in Eq. (9.4) was constructed by defining the necessary variables and functions and the solution of the differential equation was obtained with the ***dsolve()*** function.

The solution of Eq. (9.4) is obtained in Programme 9.6. We want to apply the initial conditions with respect to the relationship between ω_0 and b. As mentioned above, there are three different cases. For simplicity and because we will follow the same steps as in the previous problem, let us start with the first case $\omega_0^2 = b^2$. This condition corresponds to the critical damping condition.

In Programme 9.7, the critical damping condition is applied in the 9th cell. In the next step, with the function ***free_Symbol(f1)*** [1] and the initial condition where the mass is located at $x = 4$ cm at $t = 0$, the constant C_1 is determined and obtained in

```
[24]: using Plots

[25]: p = plot(X(t), 0, 2pi, label = "X(t) : $X(t)", title = "Simple harmonic", legend=:outertopright,
          size=(1000, 400), margin=10Plots.mm)
      plot!(U, 0, 2pi, label = "U(t): $U(t)")
      plot!(T, 0, 2pi, label = "T(t) : $T(t)")
      plot!(simplify(T + U), 0, 2pi, label = "Total energy")
      display(p)
```

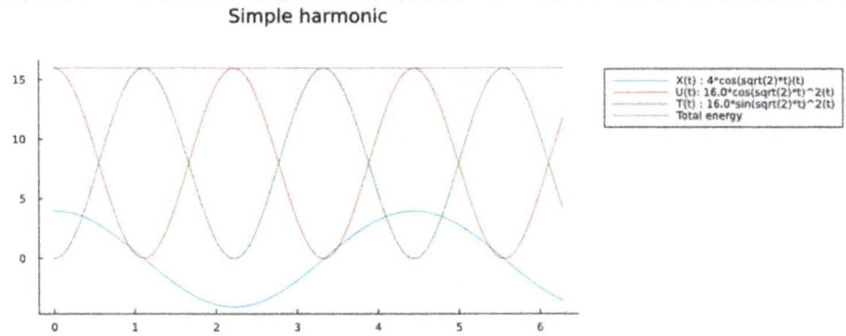

Programme 9.5 All graphics of the mass-spring system absence of friction

```
[2]: using SymPy

[3]: t = symbols(:t, real = true, positive = true)
     b = symbols(:b)
     ω0 = symbols(:ω0)
     x = SymFunction("x")(t)
```

[3]: $x(t)$

```
[4]: diff_eq = Eq(diff(x(t), t, t), - ω0^2 * x - 2 * b * diff(x(t), t))
```

[4]: $\frac{d^2}{dt^2}x(t) = -2b\frac{d}{dt}x(t) - \omega_0^2 x(t)$

```
[5]: sol = dsolve(diff_eq, x(t))
```

[5]: $x(t) = C_1 e^{t\left(-b+\sqrt{b^2-\omega_0^2}\right)} + C_2 e^{-t\left(b+\sqrt{b^2-\omega_0^2}\right)}$

```
[6]: simplify(sol)
```

[6]: $x(t) = \left(C_1 e^{t\sqrt{b^2-\omega_0^2}} + C_2 e^{-t\sqrt{b^2-\omega_0^2}}\right)e^{-bt}$

Programme 9.6 The differential equation for damped oscillation

terms of C_2. In the next step, the ***f_critical_damping*** solution was obtained by substituting the value of C_1 in the solution expression. In the 14th cell of the programme, the solution was obtained by taking values of ***10*** and ***1*** for the spring constant and mass respectively. Now that the solution is known, we can draw a graph for the critically damped case (Programme 9.8) by using Plot module [4].

Now we want to examine the over-critical and sub-critical damping conditions. For this purpose, we first want to determine the integral constants for the initial conditions. Similarly, using the function ***free_symbols(fa)*** [1], we want to first find C_1 and then find the value of C_2 using the initial value of the velocity being 0 at

```
[8]: f1 = simplify(sol.rhs())
```

[8]: $\left(C_1 e^{t\sqrt{b^2-\omega_0^2}} + C_2 e^{-t\sqrt{b^2-\omega_0^2}}\right) e^{-bt}$

```
[9]: f2 = subs(f1, Dict((b => ω0)))
```

[9]: $(C_1 + C_2)e^{-t\omega_0}$

```
[10]: c1 = free_symbols(f1)[1]
```

[10]: C_1

```
[11]: c1v = solve(f1(t => 0) ~ 4, c1)[1]
```

[11]: $4 - C_2$

```
[12]: f_critical_damping = f2(c1 => Sym(c1v[1]))
```

[12]: $4e^{-t\omega_0}$

```
[13]: f_critical_damping
```

[13]: $4e^{-t\omega_0}$

```
[14]: m, k = symbols(:m), symbols(:k)

      f_critical_damping = subs(f_critical_damping, (ω0, sqrt(k/m)), Dict((k => 10), (m => 1)))
```

[14]: $4e^{-\sqrt{10}t}$

Programme 9.7 The solution for the critical damping

$t = 0$. Then we will compare the solution we will obtain for ω_0 and b to obtain the sub-critical and over-damping states (Programmes 9.9, 9.10 and 9.11).

The next step is the sub-critical damping case and in fact the solution was obtained in the previous programme. All that remains to be done is to determine the relationship between the damping parameter and the term corresponding to the angular frequency

```
[15]: using Plots

[16]: p = plot(f_critical_damping, 0, 2, label = "X(t): $f_critical_damping", title = "Damped Oscillation",
          xlabel = "t", ylabel = "x(t)", legend =:outertopright, size = (800, 400), margin = 10Plots.mm)
```

[16]:

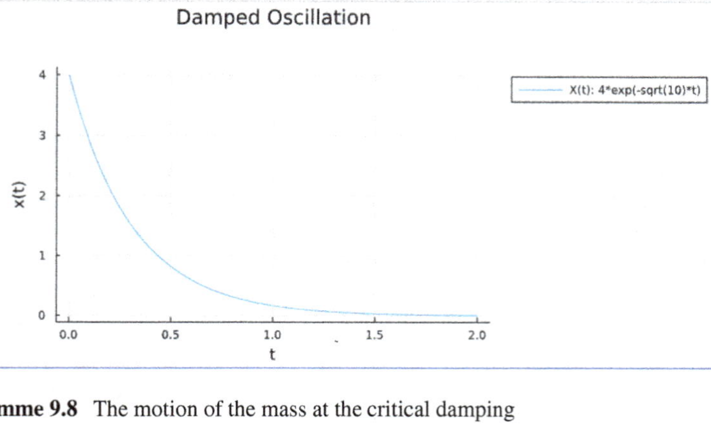

Programme 9.8 The motion of the mass at the critical damping

```
[18]: sol
```

[18]: $$x(t) = C_1 e^{t\left(-b+\sqrt{b^2-\omega_0^2}\right)} + C_2 e^{-t\left(b+\sqrt{b^2-\omega_0^2}\right)}$$

```
[19]: fa = sol.rhs()
```

[19]: $$C_1 e^{t\left(-b+\sqrt{b^2-\omega_0^2}\right)} + C_2 e^{-t\left(b+\sqrt{b^2-\omega_0^2}\right)}$$

```
[20]: c1 = free_symbols(fa)[1]
```

[20]: C_1

```
[21]: c1v = solve(fa(t => 0) ~ 4, c1)[1]
```

[21]: $4 - C_2$

Programme 9.9 Finding the first integral constant

```
[22]: fa = fa(c1 => Sym(c1v[1]))
```

[22]: $$C_2 e^{-t\left(b+\sqrt{b^2-\omega_0^2}\right)} + (4 - C_2) e^{t\left(-b+\sqrt{b^2-\omega_0^2}\right)}$$

```
[23]: C2 = free_symbols(fa)[1]
```

[23]: C_2

```
[24]: C2v = solve(diff(fa, t)(t => 0), free_symbols(fa)[1])
```

[24]: $$\left[-\frac{2b}{\sqrt{b^2-\omega_0^2}}+2\right]$$

```
[25]: fa = fa(C2 => Sym(C2v[1]))
```

[25]: $$\left(-\frac{2b}{\sqrt{b^2-\omega_0^2}}+2\right)e^{-t\left(b+\sqrt{b^2-\omega_0^2}\right)} + \left(\frac{2b}{\sqrt{b^2-\omega_0^2}}+2\right)e^{t\left(-b+\sqrt{b^2-\omega_0^2}\right)}$$

```
[26]: ft = fa
```

[26]: $$\left(-\frac{2b}{\sqrt{b^2-\omega_0^2}}+2\right)e^{-t\left(b+\sqrt{b^2-\omega_0^2}\right)} + \left(\frac{2b}{\sqrt{b^2-\omega_0^2}}+2\right)e^{t\left(-b+\sqrt{b^2-\omega_0^2}\right)}$$

```
[27]: fb = simplify(fa)
```

[27]: $$\frac{2\left(\left(-b+\sqrt{b^2-\omega_0^2}\right)e^{t\left(b-\sqrt{b^2-\omega_0^2}\right)} + \left(b+\sqrt{b^2-\omega_0^2}\right)e^{t\left(b+\sqrt{b^2-\omega_0^2}\right)}\right)e^{-2bt}}{\sqrt{b^2-\omega_0^2}}$$

```
[28]: m, k = symbols(:m), symbols(:k)

      f_over_damping = factor(subs(fa, (w0, sqrt(k/m)), Dict((k => 10), (m => 1), (b => 10))))
```

[28]: $$\frac{2\cdot\left(3e^{6\sqrt{10}t}+\sqrt{10}e^{6\sqrt{10}t}-\sqrt{10}+3\right)e^{-10t}e^{-3\sqrt{10}t}}{3}$$

Programme 9.10 Finding the solution of the differential equation according to the initial conditions

in the absence of damping. This is done in Programme 9.12 and all three possible cases are plotted together.

Programme 9.12 uses a feature that has not been mentioned before. If you pay attention to the 36th cell of the programme, ***.expand(complex = true)*** property is used. This expression applies the Euler expansion and helps to determine the real

```
[29]: plot!(f_over_damping, 0, 2, label = "Over Damping",
          legend =:outertopright, size = (800, 400), margin = 10Plots.mm)
```

[29]:

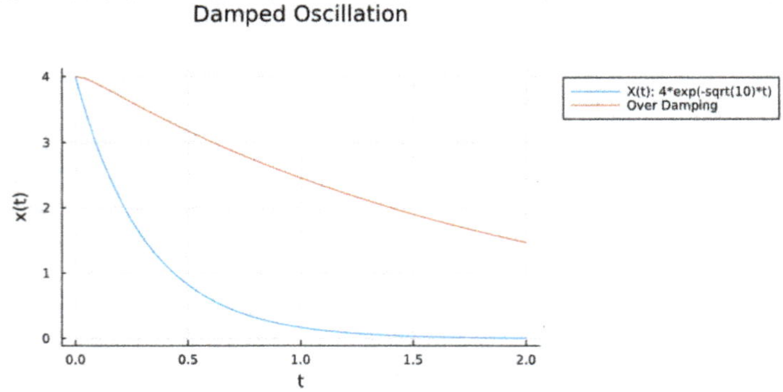

Programme 9.11 Over damping and the critical damping

```
[34]: ft = simplify(ft)
```

[34]:

$$\frac{2\left(\left(-b+\sqrt{b^2-\omega_0^2}\right)e^{t\left(b-\sqrt{b^2-\omega_0^2}\right)}+\left(b+\sqrt{b^2-\omega_0^2}\right)e^{t\left(b+\sqrt{b^2-\omega_0^2}\right)}\right)e^{-2bt}}{\sqrt{b^2-\omega_0^2}}$$

```
[35]: f_under_damping = subs(ft, Dict((b => 1), (ω0 => 9)))
```

[35]:

$$-\frac{\sqrt{5}i\left(\left(-1+4\sqrt{5}i\right)e^{t\left(1-4\sqrt{5}i\right)}+\left(1+4\sqrt{5}i\right)e^{t\left(1+4\sqrt{5}i\right)}\right)e^{-2t}}{10}$$

```
[36]: t = symbols(:t, real = true)

      f_under_damping.expand(complex = true)
```

[36]:

$$\frac{\sqrt{5}e^{-t}\sin\left(4\sqrt{5}t\right)}{5}+4e^{-t}\cos\left(4\sqrt{5}t\right)$$

```
[37]: plot!(f_under_damping, 0, 2, label = "Under Damping",
          legend =:outertopright, size = (800, 400), margin = 10Plots.mm)
```

[37]:

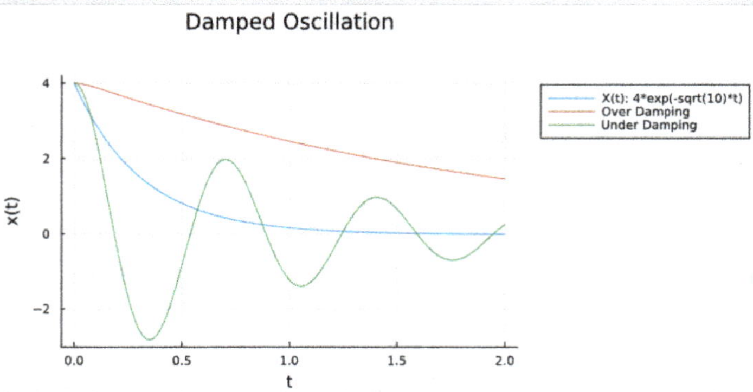

Programme 9.12 All possible situations for damped oscillation

```
[32]: u = symbols(:u, real = true, positive = true)
[32]: u
[33]: expanded_func = exp(im * u).expand(complex = true)
[33]: i sin (u) + cos (u)
```

Programme 9.13 The application of the *.expand(complex = true)*

parts of the expression. Thus, the graph in Programme 9.12 can be obtained. A simple application of the ***.expand(complex = true)*** property is given in Programme 9.13.

9.4 Two-Dimensional Oscillator

Under this title, we want to analyse the trajectory of the mass under ideal conditions such as no friction for a two-dimensional oscillator. The trajectories of the mass for such an oscillator are closely related to the phase difference and are called Lissajous curves. These curves can also be observed on the oscilloscope screen using both channels of an oscilloscope.

Here we consider a mass on a plane connected by two springs in the direction of the x and y axis. Equation of motion of the mass is given by

$$\vec{F} = -k\vec{r}. \tag{9.5}$$

If we consider this equation separated into its components, then we obtain the following two equations and their solutions:

$$F_x = -kr\cos\theta = -kx, \tag{9.6}$$

$$F_y = -kr\sin\theta = -ky. \tag{9.7}$$

The solutions of these equations are

$$x(t) = A\cos(\omega_0 t - \alpha), \tag{9.8}$$

$$y(t) = B\sin(\omega_0 t - \beta). \tag{9.9}$$

Equations (9.8 and 9.9) are sufficient for us to plot the trajectory of the oscillator. However, there are some important cases in this section, and we recommend referring to Thornton and Marion [1].

Now, let's start with loading our *SymPy* module and defining our variables. We will not write very long codes here, we will only change the phase differences for x

and y, assuming the amplitudes and angular frequencies are equal. Programme 9.14 shows the definition of the arguments and the functions given by Eqs. (9.8 and 9.9).

After these definitions, all we need to do is to assign values to the amplitudes, angular frequencies and phase differences with the ***subs()*** function. When the phase difference is 0, a line equation changing in a direct proportional way as in Programme 9.15 is obtained and when the phase difference is −pi, a trajectory changing in an inverse proportional way as in Programme 9.16 is obtained. If the phase difference is 90°, then a full circle is obtained. The mass moves on this circular trajectory exactly as given in Programme 9.17.

In the above cases we have considered a very special case where both the amplitudes and angular velocities are equal. But what if they were not? Please refer to Thornton's book and try to generalise the cases given here (Programmes 9.18 and 9.19).

```
[1]: using SymPy, Plots

[2]: t = symbols(:t, real = true, positive = true)
     a, b = symbols(:a), symbols(:b)
     w0 = symbols(:w0)
     A, B = symbols(:A), symbols(:B)

     x = SymFunction("x")(t)
     y = SymFunction("y")(t)

     x = A * cos(w0 * t - a)
```

[2]: $A \cos(a - tw_0)$

```
[3]: y = B * cos(w0 * t - b)
```

[3]: $B \cos(b - tw_0)$

Programme 9.14 The definitions of variables and the functions for two-dimensional oscillator

```
[4]: x1 = subs(x, Dict((A => 1), (w0 => 4), (a => pi/2)))
     y1 = subs(y, Dict((B => 1), (w0 => 4), (b => pi/2)))

     p = plot(x1, y1, -2, 2, label = "phase difference : $(pi/2-pi/2)", xlabel = "x", ylabel = "y",
         title = "Trajectory of oscillator", size = (300, 300), legend = :top)
     display(p)
```

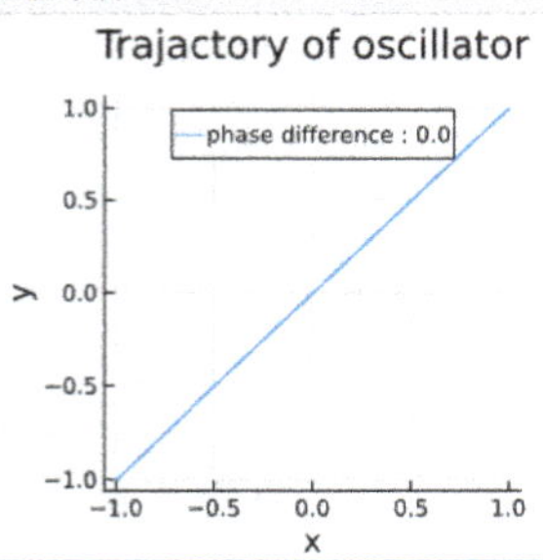

Programme 9.15 The trajectory for $a - b = 0$

```
[5]: x2 = subs(x, Dict((A => 1), (w0 => 4), (a => -pi/2)))
     y2 = subs(y, Dict((B => 1), (w0 => 4), (b => pi/2)))

     p2 = plot(x2, y2, -2, 2, label = "phase difference : -π",  xlabel = "x", ylabel = "y",
         title = "Trajactory of oscillator", size = (300, 300), legend = :top)
     display(p2)
```

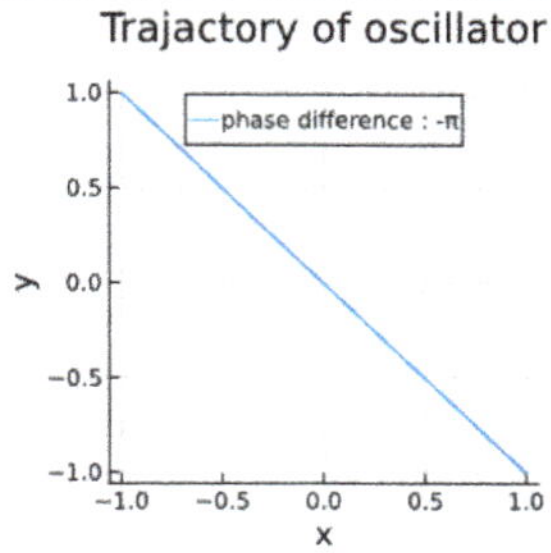

Programme 9.16 The trajectory for $a - b = -pi$

```
[6]: x3 = subs(x, Dict((A => 1), (w0 => 4), (a => 0)))
     y3 = subs(y, Dict((B => 1), (w0 => 4), (b => pi/2)))

     p3 = plot(x3, y3, -2, 2, label = "phase difference : -π/2",  xlabel = "x", ylabel = "y",
         title = "Trajactory of oscillator", size = (300, 300), legend = :right)
     display(p3)
```

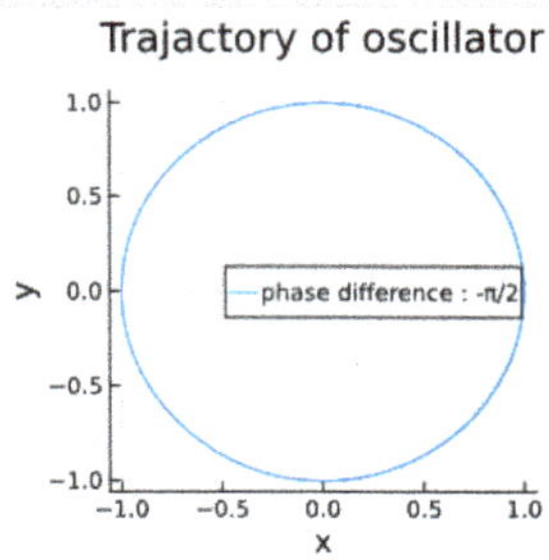

Programme 9.17 The trajectory for $a - b = pi/2$

Programme 9.20 is a code fragment that draws the trajectories collectively by travelling over the phase difference defined by both a and b with the for loop and is accompanied by the trajectories generated with this code fragment.

In the above investigations, the angular frequencies of the two-dimensional oscillator were taken equal. But this is not mandatory. The angular frequencies can be different. In particular, if the ratio of the angular frequencies is a rational number, then the trajectory of the oscillator is closed, but if the ratio of the angular frequencies is not a rational number then the trajectory is an open curve and if one waits long enough the oscillator will move to fill a 2A × 2B rectangle (Programmes 9.21 and 9.22).

```
[7]: x4 = subs(x, Dict((A => 1), (w0 => 4), (a => pi)))
     y4 = subs(y, Dict((B => 1), (w0 => 4), (b => pi/10)))

     p4 = plot(x4, y4, -2, 2, label = "phase difference : 9π/10",  xlabel = "x", ylabel = "y",
         title = "Trajectory of oscillator", size = (300, 300), legend = :right)
     display(p4)
```

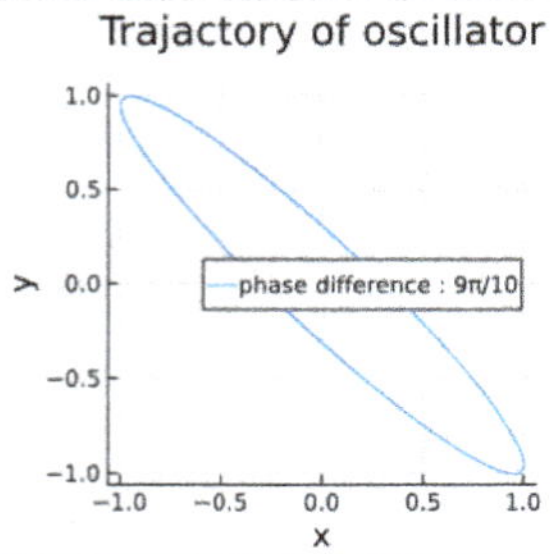

Programme 9.18 The trajectory for $a - b = 9pi/10$

```
[6]: x5 = subs(x, Dict((A => 1), (w0 => 4), (a => 0)))
     y5 = subs(y, Dict((B => 2), (w0 => 4), (b => pi/3)))

     p5 = plot(x5, y5, -1, 1, label = "phase difference : -π/3", xlabel = "x", ylabel = "y",
         title = "Trajectory of oscillator", size = (300, 300), legend = :right)
     display(p5)
```

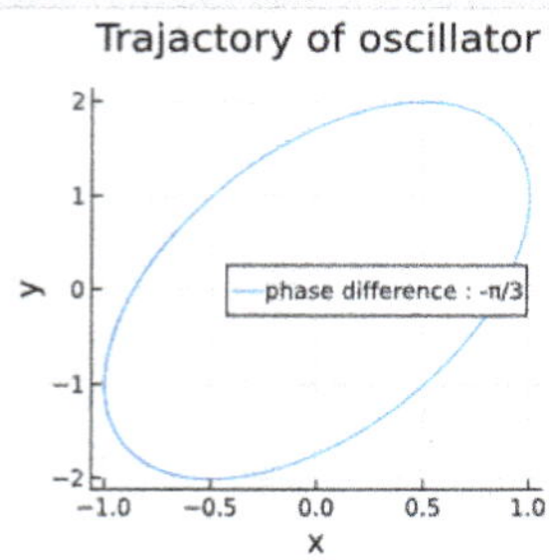

Programme 9.19 The trajectory for $a - b = -pi/3$

9.5 Mass-Spring System with Lagrange Mechanics

This problem has already been analysed using Newton's second law and Hooke's law. Here we consider the same problem with the Lagrange formalism. As it is known, the Lagrange function is an equation derived from the Euler equation, which is referred to as the Euler-Lagrange equation in some sources. The Lagrange function is the difference of kinetic and potential energy written in generalised coordinates of the system. This function must satisfy the Euler equation. Here we consider a one-dimensional harmonic oscillator. Therefore, the system has 1 degree of freedom, and the only generalised coordinate is x. The kinetic energy is given by

```
[10]: function lissajous_curve(a, b, w0, t_range)
          x = [sin(w0 * t + a) for t in t_range]
          y = [sin(w0 * t + b) for t in t_range]
          return x, y
      end
      function lissajous_plot(a_values, b_values, w0 = 4, t_range = 0:0.01:2π)
          plot()
          for a in a_values
              for b in b_values
                  x, y = lissajous_curve(a, b, w0, t_range)
                  plot!(x, y, legend =false, size = (400, 400))
              end
          end
          xlabel!("x")
          ylabel!("y")
          title!("Trajectories of oscillator")
      end
      a_values = [0, π/6, π/3, π/4, π/2, π]
      b_values = [0, π/16, π/3, π/4, π/2, π]
      lissajous_plot(a_values, b_values)
```

[10]:

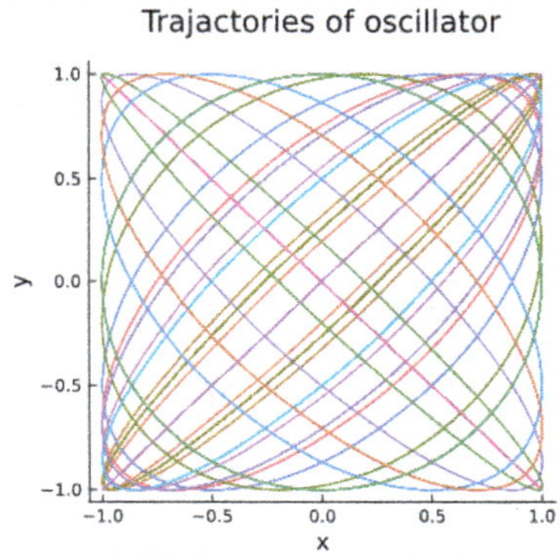

Programme 9.20 The trajectories for varying phase differences

```
[29]: x1 = subs(x, Dict((A => 1), (w0 => 2), (a => pi/2)))
      y1 = subs(y, Dict((B => 1), (w0 => 4), (b => pi/2)))

      p = plot(x1, y1, -2, 2, label = "phase difference : $(pi/2-pi/2)", xlabel = "x", ylabel = "y",
          title = "Lissajous curve", size = (300, 300), legend = :top)
      display(p)
```

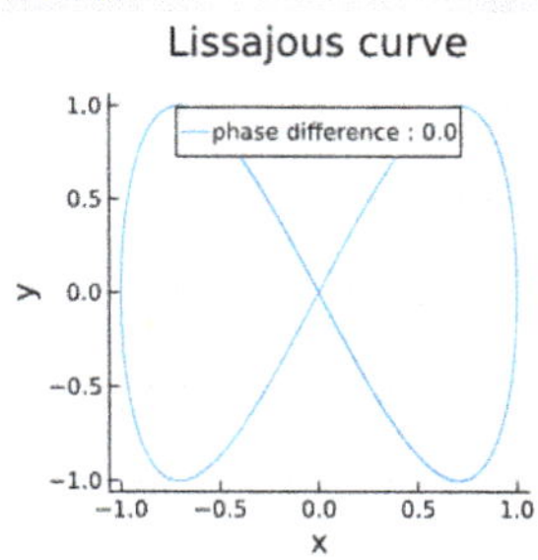

Programme 9.21 The Lissajous curve for different angular frequencies

```
[31]: x1 = subs(x, Dict((A => 1), (w0 => 1), (a => pi/2)))
      y1 = subs(y, Dict((B => 1), (w0 => 4), (b => pi/2)))

      p = plot(x1, y1, -4, 4, label = "phase difference : $(pi/2-pi/2)",  xlabel = "x", ylabel = "y",
          title = "Lissajous curve", size = (300, 300), legend = :top)
      display(p)
```

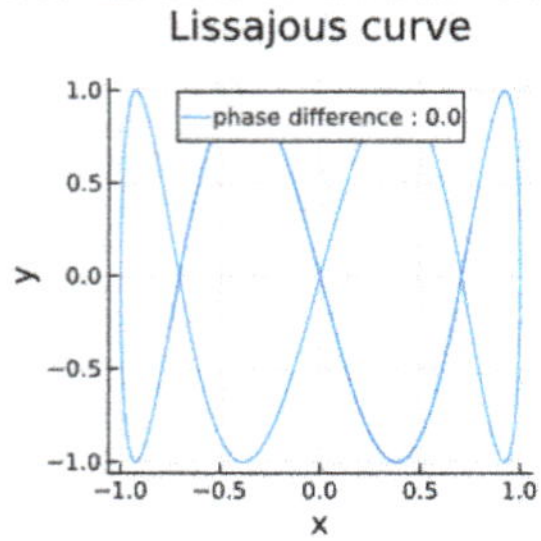

Programme 9.22 The Lissajous curve for different angular frequencies

$$T = \frac{1}{2}m\left(\frac{d}{dt}x(t)\right)^2 = \frac{1}{2}m\dot{x}(t)^2, \tag{9.10}$$

and the potential energy is given by

$$U = \frac{1}{2}kx(t)^2. \tag{9.11}$$

Therefore, the Lagrange function equals to

$$L = T - U = \frac{1}{2}m\dot{x}(t)^2 - \frac{1}{2}kx(t)^2 \tag{9.12}$$

The Lagrange function must satisfy the Euler equation, where we are dealing with a system with one degree of freedom and the generalised coordinate x:

$$\frac{\partial L}{\partial x} - \frac{d}{dt}\left(\frac{\partial L}{\partial \dot{x}}\right) = 0. \tag{9.13}$$

Now we want to write and solve these equations in Julia and again using the *SymPy* module. After the *SymPy* module is added to the programme, m, k and the time variable t are defined as symbolic variables in Programme 9.23. At the same time, kinetic energy ***T*** and potential energy ***U*** are defined as functions with their dependent variables (Programme 9.24).

The result obtained in Programme 9.25 is in terms of trigonometric functions rather than exponential functions. The reason for this is explained in detail in the variables defined in Sect. 9.2 of this chapter and the variables defined here. Initial values will not be applied to the solution obtained in Programme 9.25 because they are explained in detail in Sect. 9.2. The problem-specific solution will be obtained

```
[1]: using SymPy

[2]: m = symbols(:m, real = true, positive = true)
     k = symbols(:k, real = true, positive = true)
     t = symbols(:t, real = true, positive = true)

     x = SymFunction("x")(t)
     T = SymFunction("T")(diff(x, t))
```

[2]: $T(\frac{d}{dt}x(t))$

```
[3]: U = SymFunction("U")(x)
```

[3]: $U(x(t))$

Programme 9.23 The definitions of the variables and the functions for the mass-spring system

```
[4]: L = SymFunction("L")
```

[4]: L

```
[5]: T = 1/2 * m * diff(x, t)^2
```

[5]: $0.5m\left(\frac{d}{dt}x(t)\right)^2$

```
[6]: U = 1/2 * k * x(t)^2
```

[6]: $0.5kx^2(t)$

```
[7]: L = T - U
```

[7]: $-0.5kx^2(t) + 0.5m\left(\frac{d}{dt}x(t)\right)^2$

Programme 9.24 The Lagrange function of the system

by following the steps in this section. In other words, the initial conditions of the problem are left to the reader.

```
[8]: F = Eq(diff(L, x), diff(diff(L, diff(x, t)), t))
```

[8]: $-1.0kx(t) = 1.0m\frac{d^2}{dt^2}x(t)$

```
[9]: f = dsolve(F, x(t))
```

[9]: $x(t) = C_1 \sin\left(\frac{\sqrt{k}t}{\sqrt{m}}\right) + C_2 \cos\left(\frac{\sqrt{k}t}{\sqrt{m}}\right)$

Programme 9.25 The application of Euler equation to the Lagrange function

9.6 Mass Rotating on a Plane with Constant Radius

Here, for example, consider a body of mass m rotating around a centre on a trajectory of constant radius on the surface of a table. We can immediately say that if the angular velocity of the mass does not change, that is, if the mass is not under the influence of an external force, the angular momentum of the mass is constant. Now let us try to analyse this problem in Julia language and again by considering the *SymPy* module.

But first we need to clarify some things. The concept of generalised coordinates in Lagrangian mechanics provides us with great convenience. The degrees of freedom of the problem are two with respect to the Cartesian coordinates defined on the table surface, which can be described by $\boldsymbol{x}$ and $\boldsymbol{y}$. If we consider the problem in Cartesian coordinates, both $\boldsymbol{x}$ and $\boldsymbol{y}$ will change together. But if we consider the problem in plane polar coordinates, we get an advantage. The degrees of freedom of the problem are still two, r and θ, but since the radius does not change during the motion of the mass, the constant radius creates a bond condition and the degrees of freedom of the system is 1 and the generalised coordinate of the problem is θ.

Kinetic energy in polar coordinates is given by

$$T=\frac{1}{2}m\left(\dot{r}^{2}\Big|_{r=const}+r^{2}\,\dot{\theta}^{2}\right) \tag{9.14}$$

And we can take the potential energy equal to zero. Therefore, the Lagrange equation is

$$L = T - U = \frac{1}{2}mr^{2}\dot{\theta}. \tag{9.15}$$

$$\frac{\partial L}{\partial \theta} - \frac{d}{dt}\left(\frac{\partial L}{\partial \dot{\theta}}\right) = 0. \tag{9.16}$$

Equation 9.16 gives that the angular momentum is conserved. Now we want to solve the problem by using Julia codes. Programme 9.26 represents the symbolic variables and the functions for the problem.

The last cell of Programme 9.27 shows that the angular momentum of the system is constant, where $\boldsymbol{a}$ is a fixed number. Here a highly idealised problem is considered, and an attempt is made to describe a procedure that can be followed for more complex cases.

With careful analytical calculations and the correct choice of generalised coordinates, much more complex problems can be realised by improving the code given here. It should be noted that, unfortunately, the choice of generalised coordinates is a matter of experience and requires a great deal of study and effort.

In this section we have tried to focus on idealised and relatively close to reality problems exhibiting periodic motions. It is intended to provide the reader with a methodisation for studying much more pleasant problems, and with sufficient care

```
[1]: using SymPy

[2]: m = symbols(:m)
     t = symbols(:t)
     a = symbols(:a)
```

[2]: a

```
[3]: r = SymFunction("r")(t)
     θ = SymFunction("θ")(t)

     T = SymFunction("T")((r), (θ), (diff(r, t)), (diff(θ, t)))
```

[3]: $T(r(t), \theta(t), \frac{d}{dt}r(t), \frac{d}{dt}\theta(t))$

```
[4]: U = SymFunction("U")(r)
```

[4]: $U(r(t))$

```
[5]: L = SymFunction("L")
```

[5]: L

Programme 9.26 The symbolic definitions for the problem

```
[6]: T = 1/2 * m * (diff(r, t)^2 + r^2 * diff(θ, t)^2)
```

[6]: $0.5m\left(r^2(t)\left(\frac{d}{dt}\theta(t)\right)^2 + \left(\frac{d}{dt}r(t)\right)^2\right)$

```
[7]: U = 0
```

[7]: 0

```
[8]: L = T - U
```

[8]: $0.5m\left(r^2(t)\left(\frac{d}{dt}\theta(t)\right)^2 + \left(\frac{d}{dt}r(t)\right)^2\right)$

```
[9]: diff(L, θ)
```

[9]: 0

```
[10]: diff(L, diff(θ, t))
```

[10]: $1.0mr^2(t)\frac{d}{dt}\theta(t)$

```
[11]: f = Eq(diff(L, diff(θ, t)), a) # where a is a constant number
```

[11]: $1.0mr^2(t)\frac{d}{dt}\theta(t) = a$

Programme 9.27 The solution of the problem

many seemingly complex problems can be made much more understandable by using appropriate Julia codes.

9.7 Problems

1. Consider a mass m suspended at the end of a rope of length ℓ. Suppose that mass $\boldsymbol{m}$ is released by making a θ angle with the vertical direction. Write the equation of motion for the mass and try to solve the equation with Julia codes without making small angle approximation. For theta, try to solve the equation of motion of the particle by rewriting it using the series expansion functions in SymPy or other symbolic computing modules.
2. In the previous problem, try to show the time dependence of potential and kinetic energy for mass mathematically and graphically.
3. For the situation in Problem 1, try to solve the equation of motion of the particle by taking into account the existence of a friction force proportional to the velocity of the mass. According to the state of the damping constant, after how many revolutions the mass is completely immobilised. You can make an approach for this and try to write Julia codes for this.
4. Write the Lagrange function for the case given in Problem 1 and solve the Lagrange equation with codes written in Julia programming language. For example, try to write a Julia function or a macro for the Lagrange equation for single degree of freedom systems or for systems that can be reduced from two degrees of freedom to a single degree of freedom with the bound condition.
5. Imagine a tunnel with one end at the north pole of the Earth and the other end at the south pole. Discuss the motion of an object released from one end of this tunnel and analyse your results in more detail with the codes you will write in the Julia program. Solve the same problem for the planets Jupyter and Mars and compare your results.
6. Physical systems in everyday life are far from idealised conditions. For example, a mass-spring system is subject to friction for a real physical system and the motion of the mass is gradually extinguished. For this purpose, forced oscillations are realised by applying a driving force to the mass. Try to determine the resonance frequency for such a system with codes written in Julia language.

References

1. Thornton, Stephen T. and Marion, Jerry B. (2004) Classical Dynamics of Particles and Systems, 6th Edition, Brooks/Cole. ISBN 0534408966, 9780534408961
2. Julia Documentation. https://julialang.org/
3. SymPy module. https://docs.sympy.org/latest/tutorials/intro-tutorial/index.html
4. Plots module. https://docs.juliaplots.org/stable/tutorial/#tutorial

Chapter 10
Julia Programming Applications in Modern Physics

10.1 Introduction

In this section, in our adventure in Quantum Mechanics, which started with processes such as understanding the duality principle for light, Planck's explanation of black body radiation by accepting light as consisting of photons with hv energy, photoelectric effect, Compton effect, Bohr's determination of the spectra of hydrogen atoms, we will try to touch upon some practical applications related to Modern Physics topics without going too far in Quantum Mechanics.

10.2 Determination of Planck's Constant with Photoelectric Data

In a photoelectric effect experiment using calcium, the values given in Table 10.1 were obtained. Let us try to determine Planck's constant by processing these data and using Einstein's equation [1].

It was first observed by Hertz that when light falls on a metal surface, electrons are removed in certain situations. In order to explain this effect and to calculate the kinetic energy of the removed electrons, Einstein simply utilised the conservation of energy. But in Einstein's equation the photon energy was taken as hv. In other words, Einstein considered that light consists of photons with energy $h\nu$.

$$T_{\max} = h\nu - \phi. \tag{10.1}$$

Here $T_{\max}$ corresponds to the maximum energy of rejected electron, h corresponds to the Planck constant, which we want to find it as numerical, ν is the frequency of the light and ϕ is the work function which corresponds to the energy required to remove electrons from the metal.

R. G. Türeci et al., *Julia Programming for Physics Applications*,
https://doi.org/10.1007/978-3-031-84716-5_10

Table 10.1 The photoelectric effect data

$\lambda(A^{\circ})$	$V_{\text{stop}}(V)$
2536	1.95
3132	0.98
3650	0,50
4047	0.14

Since it is somewhat difficult to measure the energy of the electrons removed in the experiment, we can apply an electric force in the opposite direction to the electrons with a stopping potential. Thus, the stopping voltage value at which the electron flow stops will give us the maximum energy of the electrons in eV. In this case Eq. (10.1) is written as follows:

$$eV_{\text{stop}} = h\nu - \phi. \tag{10.2}$$

$$V_{\text{stop}} = \frac{h}{e}\nu - \frac{\phi}{e} \tag{10.3}$$

Let's start with transferring the data in Table 10.1 to a DataFrame. In doing so, however, we want to convert the wavelength of the light used into frequency. In this DataFrame, if we take the frequency as the independent variable and the stop voltage as the dependent variable and plot the data, we obtain a line equation. The slope of this line will give us the ***h/e*** value. Therefore, by multiplying the slope by the electron charge, we determine the value of Planck's constant (Programme 10.1).

The desired result is obtained in Programme 10.2. In fact, the slope of the line could have been found directly by applying linear regression, which is a machine learning technique. The interested reader can make such a calculation as an application.

10.3 Wave Packet Concept

In quantum mechanics, particles are defined as matter waves and are the third wave type in physics after mechanical waves and electromagnetic waves. Let us consider a simple example from many modern physics textbooks of how a particle can be thought of as a matter wave. Consider the case where two sinusoidal waves of close frequency are superimposed on each other.

$$y_1 = A\sin(\omega_1 t) \tag{10.4}$$

$$y_2 = A\sin(\omega_2 t) \tag{10.5}$$

The total wave after some algebraic manipulations is given by

[1]:
```
using DataFrames, Plots
```

[2]:
```
df = DataFrame()
df.wavelength = [2536, 3132, 3650, 4047]
df.frequency = (3 * 10^18) ./ df.wavelength
df.stop = [1.95, 0.98, 0.50, 0.14]

show(df, eltypes = false)
```

```
4×3 DataFrame
 Row │ wavelength  frequency   stop
─────┼──────────────────────────────
   1 │       2536  1.18297e15  1.95
   2 │       3132  9.57854e14  0.98
   3 │       3650  8.21918e14  0.5
   4 │       4047  7.4129e14   0.14
```

[3]:
```
p = scatter(df.frequency, df.stop, label = "photoelectric effect", size = (400, 250))
plot!(df.frequency, df.stop, label = "photoelectric effect")
```

[3]:

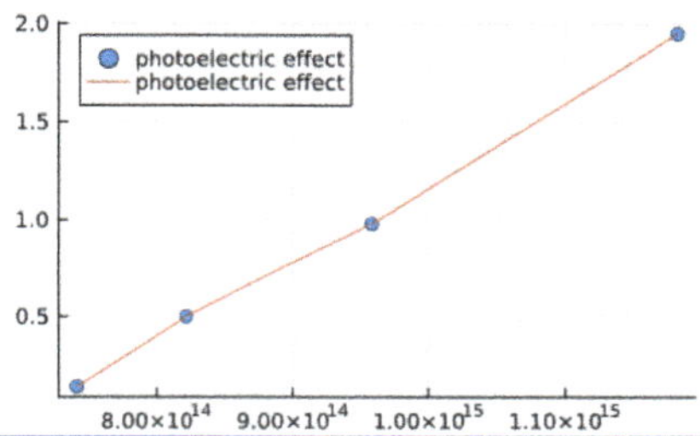

Programme 10.1 The DataFrame prepared with the data in Table 10.1

[4]:
```
slope = (df.stop[4] - df.stop[2]) / (df.frequency[4] - df.frequency[2])
h_value = slope * 1.6 * 10^(-19)
```

[4]:
```
6.206001521311476e-34
```

Programme 10.2 The value of Planck constant

$$y = y_1 + y_2 = A\cos\left(\frac{\omega_1 - \omega_2}{2}t\right)\sin\left(\frac{\omega_1 + \omega_2}{2}t\right) = A(t)\sin\left(\frac{\omega_1 + \omega_2}{2}t\right) \quad (10.6)$$

In Eq. 10.6, the cosine term on the right-hand side of the equation repeats itself with a lower frequency, while the sine term repeats itself with a higher frequency and these two are actually a single wave. What we want to show here will be clearer after plotting the wave with the Julia programme. The term oscillating with the lower frequency is called the envelope of the wave and a wave packet is obtained as shown in the Programme 10.3.

```
[1]: using SymPy, Plots

[2]: t = symbols(:t, real = true, positive = true)
     A = symbols(:A)
     w1, w2 = symbols(:w1), symbols(:w2)

     y1 = A * sin.(w1 .* t)
     y2 = A * sin.(w2 .* t)
```

[2]: $A \sin(t w_2)$

```
[3]: f1 = subs(y1, Dict((A => 1), (w1 => 2pi)))
     f2 = subs(y2, Dict((A => 1), (w2 => 2pi + 1.5)))
     f = f1 + f2

     envelope = 2 * A * (cos((w1 - w2) * t / 2))
     fe = subs(envelope, Dict((A => 1), (w1 => 2pi), (w2 => 2pi + 1.5)))

     p = plot(t, f, 0, 3pi, label="Wave Function", linewidth = 2, size = (500, 300))
     plot!(t, fe, 0, 3pi, label="Envelope:$fe", linewidth = 1.5)
```

[3]:

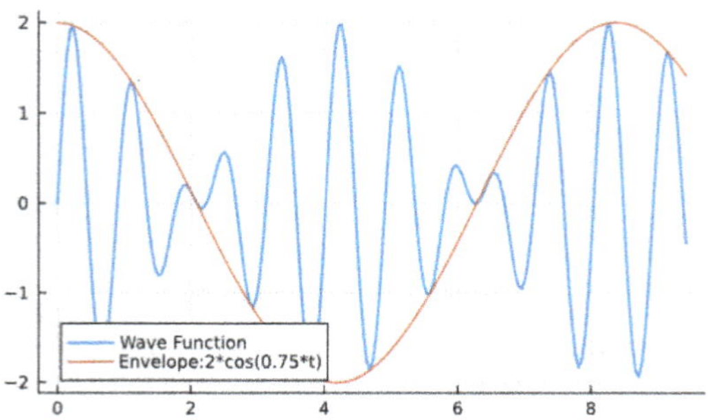

Programme 10.3 The wave packet

10.4 Relationship Between Phase and Group Velocity and Dispersion Properties of a Medium

The phase velocity gives the propagation speed of the phase in a wave packet. For example, the speed of the peak or trough of a wave. The phase velocity is determined by the ratio of the angular velocity of the wave to the number of waves, $\upsilon_p = \omega / k$. In contrast, the group velocity gives the propagation speed of the entire wave packet, which is the rate at which energy or information is transmitted. The group velocity is given by $\upsilon_g = d\omega / dk$.

The speed of a wave travelling in a medium with refractive index n is given by

$$\upsilon_g = \upsilon_p - \lambda \frac{d\upsilon_p}{d\lambda}. \tag{10.7}$$

This relation gives information about the dispersion property of the medium through which the wave travels [5]. For example, both transverse and longitudinal waves can propagate in solids. In fluids, on the other hand, longitudinal waves usually propagate. In addition, in some fluids such as water, transverse ripple waves with

```
[1]: using SymPy

     λ, ρ, T = symbols(:λ), symbols(:ρ), symbols(:T)
     Vp = SymFunction("Vp")(λ)
     Vg = SymFunction("Vg")(λ)

     Vp = sqrt(2pi * T / (ρ * λ))

     Vg = Vp - λ * diff(Vp, λ)
```

[1]: $3.7599424119465\sqrt{\frac{T}{\lambda\rho}}$

```
[2]: Vg / Vp

     if Vg / Vp > 1
         println("the medium is an anormal dispersion, Vg/Vp : $(Vg/Vp)")
     else
         println("the medium is an normal dispersion, Vg/Vp : $(Vg/Vp)")
     end

     the medium is an anormal dispersion, Vg/Vp : 1.50000000000000
```

Programme 10.4 The dispersion relation calculation

$\lambda \leq 1$ cm can occur due to surface tension. The phase velocity of these small ripple waves is given by

$$\upsilon_p = \sqrt{\frac{2\pi T}{\rho\lambda}}, \tag{10.8}$$

where ρ is the density of the liquid, T is surface tension and λ is the wavelength of the wave. Now let us try to find the group velocity of these transverse ripple waves and determine the dispersion property of water for these ripple waves. The calculation is given in Programme 10.4. Programme 10.5 shows the variation in both phase velocity and group velocity.

10.5 Particle Interacting with a Potential Barrier

Here we want to study the case of a particle approaching and interacting with a uniform potential barrier of height U_0 and width in one dimension with energy $E < U_0$. From the point of view of classical physics, since the energy of the particle is smaller than the height of the obstacle, the particle reflects backwards from the obstacle. We also want to analyse this problem quantum mechanically [1]. The potential barrier is given in Fig. 10.1.

In order to analyse the problem, we need to write the Schrödinger equation for all three regions. In $x < 0$ and $x > a$ the energy of the particle is E, but in $0 \leq x \leq a$ the energy of the particle is $U_0 - E$.

Schrödinger equations are given

```
[4]: Vpv = subs(Vp, Dict((T => 1), (ρ => 1.2)))
     Vgv = subs(Vg, Dict((T => 1), (ρ => 1.2)))

     using Plots
     λ = 0:100:4000
     plot(λ, Vpv, label = "Phase velocity", xlabel = "λ",
         ylabel = "phase and grouup velocity")
     plot!(λ, Vgv, label = "Group velocity")
```

[4]:

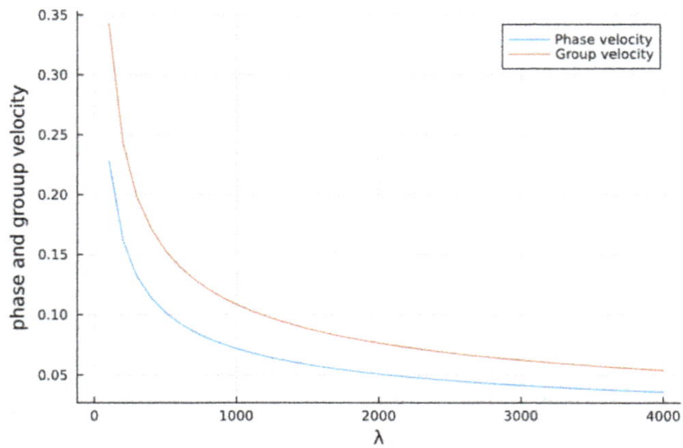

Programme 10.5 The variation of the phase and group velocities for the phase velocity given in Eq. (10.8)

Fig. 10.1 The potential barrier

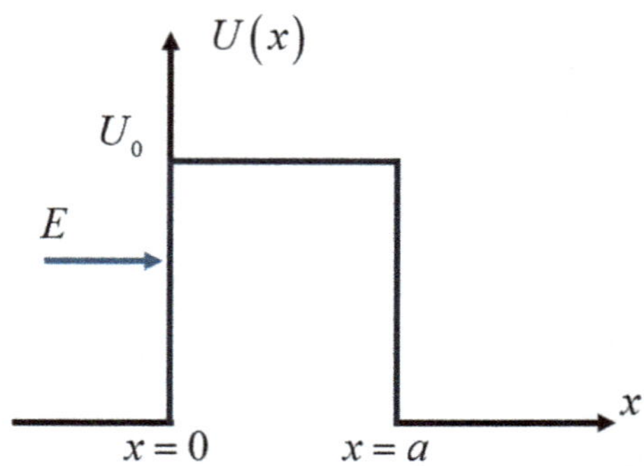

$$-\frac{\hbar^2}{2m}\frac{d^2\psi_1(x)}{dx^2} = E\psi_1(x) \text{ for } x < 0, \tag{10.9}$$

$$-\frac{\hbar^2}{2m}\frac{d^2\psi_2(x)}{dx^2} + U_0\psi_2(x) = E\psi_2(x) \text{ for } 0 \le x \le a, \tag{10.10}$$

$$-\frac{\hbar^2}{2m}\frac{d^2\psi_3(x)}{dx^2} = E\psi_3(x) \text{ for } x > a, \tag{10.11}$$

```
[1]: using SymPy, Plots, LinearAlgebra

[2]: Ai = symbols(:Ai, real = true, positive = true)
     Ay = symbols(:Ay, real = true, positive = true)
     C = symbols(:C, real = true, positive = true)
     D = symbols(:D, real = true, positive = true)
     Ag = symbols(:T, real = true, positive = true)
     a = symbols(:a, real = true, positive = true)
     k = symbols(:k, real = true, positive = true)
     q = symbols(:q, real = true, positive = true)
```

[2]: q

```
[3]: f1 = Eq(Ai + Ay - C - D, 0)
```

[3]: $Ai + Ay - C - D = 0$

```
[4]: f2 = Eq(im*k*Ai - im*k*Ay + q*C - q*D, 0)
```

[4]: $iAik - iAyk + Cq - Dq = 0$

```
[5]: f3 = Eq(0*Ai + 0*Ay + C*exp(-q*a) + D*exp(q*a), Ag*exp(im*k*a))
```

[5]: $Ce^{-aq} + De^{aq} = Te^{iak}$

```
[6]: f4 = Eq(0*Ai + 0*Ay - q*C*exp(-q*a) + q*D*exp(q * a), im*k*Ag*exp(im*k*a))
```

[6]: $-Cqe^{-aq} + Dqe^{aq} = iTke^{iak}$

Programme 10.6 Importing *SymPy* module and the variables

for each region respectively. These equations are similar to the equation which was studied in the previous section for the mass-spring system. We can perform the following definitions

$$k^2 = \frac{2mE}{\hbar^3}, \tag{10.12}$$

$$q^2 = \frac{2m(U_0 - E)}{\hbar^3}. \tag{10.13}$$

The solutions of Eqs. (10.9)–(10.11) are given by

$$\psi_1(x) = A_{in}\exp(ikx) + A_{ref}\exp(-ikx), \tag{10.14}$$

$$\psi_2(x) = C\exp(qx) + D\exp(-qx), \tag{10.15}$$

$$\psi_3(x) = A_{tr}\exp(ikx) + \cancel{A_{ref}}_0\exp(-ikx). \tag{10.16}$$

Since there is no potential barrier in the $x > a$ region, there will be no reflected wave part in this region. Therefore, we have taken the coefficient of the second term on the right-hand side of Eq. (10.16) as zero. For the problem we have the following boundary conditions such that both the wave functions and the first derivatives of the wave functions must be equal at $x = 0$ and $x = a$. Therefore, our boundary conditions are as follows:

$$\psi_1(x)|_{x=0} = \psi_2(x)|_{x=0} \qquad \psi_2(x)|_{x=a} = \psi_3(x)|_{x=a} \tag{10.17}$$

$$\left.\frac{d\psi_1(x)}{dx}\right|_{x=0} = \left.\frac{d\psi_2(x)}{dx}\right|_{x=0} \qquad \left.\frac{d\psi_2(x)}{dx}\right|_{x=a} = \left.\frac{d\psi_3(x)}{dx}\right|_{x=a} \tag{10.18}$$

By applying the boundary conditions given in Eqs. (10.17) and (10.18), the following equations are obtained.

$$A_{in} + A_{ref} = C + D \tag{10.19}$$

$$ikA_{in} - ikA_{ref} = -qC + qD \tag{10.20}$$

$$Ce^{-qa} + De^{qa} = A_{tr}e^{ika} \tag{10.21}$$

$$-qCe^{-qa} + qDe^{qa} = ikA_{tr}e^{ika} \tag{10.22}$$

To determine these coefficients, these four equations must be solved together. A practical way to do this is to use the matrix formalism. We can write the coefficients of the equation we want to solve in matrix notation as follows.

We want to solve this matrix equation using Julia codes. Our aim is to determine the coefficients and by looking at the ratios of these coefficients, we want to find the ratio of the particle passing through the obstacle and reflecting from the barrier.

$$\begin{bmatrix} 1 & 1 & -1 & -1 \\ ik & -ik & q & -q \\ 0 & 0 & \exp(-qa) & \exp(qa) \\ 0 & 0 & -q\exp(-qa) & q\exp(qa) \end{bmatrix} \cdot \begin{bmatrix} A_{in} \\ A_{ref} \\ C \\ D \end{bmatrix} = \begin{bmatrix} 0 \\ 0 \\ A_{tr}\exp(ika) \\ ikA_{tr}\exp(ika) \end{bmatrix} \tag{10.23}$$

$$\alpha_{12} = \frac{|C|^2}{|A_{\mathrm{in}}|^2}, \tag{10.24}$$

$$\alpha_{13} = \frac{|A_{tr}|^2}{|A_{\mathrm{in}}|^2}, \tag{10.25}$$

$$\beta_{12} = \frac{|A_{\mathrm{ref}}|^2}{|A_{\mathrm{in}}|^2}, \tag{10.26}$$

$$\beta_{23} = \frac{|D|^2}{|C|^2}. \tag{10.27}$$

```
[7]: A = [1 1 -1 -1;
          im*k -im*k q -q;
          0 0 exp(-q*a) exp(q*a);
          0 0 -q*exp(-q*a) q*exp(q*a)]
```

[7]:

$$\begin{bmatrix} 1 & 1 & -1 & -1 \\ ik & -ik & q & -q \\ 0 & 0 & e^{-aq} & e^{aq} \\ 0 & 0 & -qe^{-aq} & qe^{aq} \end{bmatrix}$$

```
[8]: B = [0, 0, Ag*exp(im*k*a), im*k*Ag*exp(im*k*a)]
```

[8]:

$$\begin{bmatrix} 0 \\ 0 \\ Te^{iak} \\ iTke^{iak} \end{bmatrix}$$

Programme 10.7 The matrix in Eq. (10.23)

Here α_{12} corresponds to the transition probability from $x < 0$ to $0 < x < a$, α_{13} corresponds to the transition probability from $x < 0$ to $x > a$. Similarly, β_{12} defines the probability of projection to $x < 0$ and β_{23} defines the probability of projection from $x = a$ to $0 < x < a$. α_{13} and β_{12} equals to the following results.

$$\alpha_{13} = \frac{1}{1 + \frac{U_0^2 \sinh^2(qa)}{4E(U_0 - E)}} \tag{10.28}$$

$$\beta_{12} = \frac{1}{1 + \frac{4E(U_0 - E)}{U_0^2 \sinh^2(qa)}} \tag{10.29}$$

Let's start with loading the *SymPy* module and defining variables.

In Programme 10.6 f1, f2, f3 and f4 are the definitions in Eqs. (10.19)–(10.22). The matrix given in Eq. (10.23) is given in Programme 10.7.

The required coefficients are calculated in Programme 10.8. Now α_{13} can be calculated. According to the mathematical formulation of quantum mechanics, the absolute value squared is the product of the coefficient and the complex conjugate of the coefficient. Therefore, the ***conj()*** function should be used to calculate the complex conjugate of the expression (Programme 10.9).

The result of Programme 10.10 is important. In terms of classical mechanics, the particle is reflected to the medium from which it came, but in terms of quantum mechanics, the probability of the particle overcoming the potential barrier is a non-zero number and this is known as tunnelling.

In this case, the sum of the probability that the particle is reflected to $x < 0$ and the probability calculated here must be equal to 1. This is the same as the sum of the probabilities of heads and tails when a coin is tossed in the air being equal to unity.

```
[9]: det_A = det(A)
```

[9]: $-4ikq$

```
[10]: A1 = copy(A)
      A1[:, 1] = B
      det_Ai = det(A1)
```

[10]: $-Tk^2e^{aq}e^{iak} + Tk^2e^{-aq}e^{iak} - 2iTkqe^{aq}e^{iak} - 2iTkqe^{-aq}e^{iak} + Tq^2e^{aq}e^{iak} - Tq^2e^{-aq}e^{iak}$

```
[11]: A2 = copy(A)
      A2[:, 2] = B
      det_Ay = det(A2)
```

[11]: $-Tk^2e^{aq}e^{iak} + Tk^2e^{-aq}e^{iak} - Tq^2e^{aq}e^{iak} + Tq^2e^{-aq}e^{iak}$

```
[12]: A3 = copy(A)
      A3[:, 3] = B
      det_C = det(A3)
```

[12]: $-2ik\left(-iTke^{aq}e^{iak} + Tqe^{aq}e^{iak}\right)$

```
[13]: A4 = copy(A)
      A4[:, 4] = B
      det_D = det(A4)
```

[13]: $-2ik\left(iTke^{-aq}e^{iak} + Tqe^{-aq}e^{iak}\right)$

Programme 10.8 The calculated coefficients

```
[14]: Agh = (2*im*k*q)*Ai /(exp(im*k*a)*((k^2-q^2)*sinh(q*a) + 2*im*k*q*cosh(q*a)))
```

[14]: $\frac{2iAikqe^{-iak}}{2ikq\cosh(aq)+(k^2-q^2)\sinh(aq)}$

```
[15]: conj(Agh)
```

[15]: $-\frac{2iAikqe^{iak}}{-2ikq\cosh(aq)+(k^2-q^2)\sinh(aq)}$

```
[16]: FF = ((Agh * conj(Agh)) / Ai^2)
```

[16]: $\frac{4k^2q^2}{(-2ikq\cosh(aq)+(k^2-q^2)\sinh(aq))(2ikq\cosh(aq)+(k^2-q^2)\sinh(aq))}$

```
[17]: simplify(FF).expand(complex = true)
```

[17]: $\frac{4k^2q^2}{k^4\sinh^2(aq)+2k^2q^2\sinh^2(aq)+4k^2q^2+q^4\sinh^2(aq)}$

Programme 10.9 The calculation of α_{13}

10.6 State of a Particle in a Two-Dimensional Infinite Potential Well

Here we want to consider the case of a particle of mass m in a two-dimensional infinite potential well. This is an interesting example worth analysing for its consequences [1]. Let us adopt the general approach that the potential well is defined in the interval $0 \leq x \leq a$ on the x-axis and $0 \leq y \leq b$ on the y-axis (Fig. 10.2).

Schrödinger equation describing the behaviour of the particle is given by

```
[20]: T = 4k^2*q^2/((k^2+q^2)^2 * sinh(q*a) + 4k^2*q^2)
```

[20]: $\frac{4k^2q^2}{4k^2q^2+(k^2+q^2)^2\sinh(aq)}$

```
[21]: U0 = symbols(:U0)
      E = symbols(:E)
      m = symbols(:m)
      h = symbols(:h)

      T1 = subs(T, Dict(((k^2 => 2*m*E/h^2) , (q^2 => 2*m*(U0-E)/h^2))))
```

[21]: $\frac{16Em^2(-E+U_0)}{h^4\cdot\left(\frac{16Em^2(-E+U_0)}{h^4}+\left(\frac{2Em}{h^2}+\frac{2m(-E+U_0)}{h^2}\right)^2\sinh\left(\sqrt{2}a\sqrt{\frac{m(-E+U_0)}{h^2}}\right)\right)}$

```
[22]: simplify(T1)
```

[22]: $\frac{4E(E-U_0)}{4E(E-U_0)-U_0^2\sinh\left(\sqrt{2}a\sqrt{-\frac{m(E-U_0)}{h^2}}\right)}$

```
[23]: # gameover ;)
```

Programme 10.10 The calculation of α_{13} in terms of the particle energy and the height of the potential barrier

Fig. 10.2 Two-dimensional infinite potential well

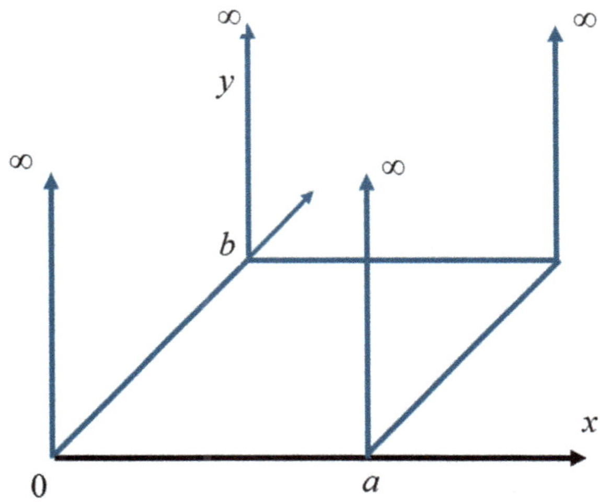

$$-\frac{\hbar^2}{2m}\left(\frac{\partial^2}{\partial x^2}+\frac{\partial^2}{\partial y^2}\right)\psi(x,y)+U(x,y)_0\psi(x,y)=E\psi(x,y). \tag{10.30}$$

If the separation of variables method is used for the wave function, for which the differential equation must be homogeneous, we obtain the following sets of equations. Then, we have two differential equations.

$$-\frac{\hbar^2}{2m}\frac{\partial^2}{\partial x^2}X(x)=E_1X(x)\Rightarrow\frac{\partial^2}{\partial x^2}X(x)=-\frac{2mE_1}{\hbar^2}X(x), \tag{10.31}$$

$$-\frac{\hbar^2}{2m}\frac{\partial^2}{\partial y^2}Y(y) = E_2 Y(y) \Rightarrow \frac{\partial^2}{\partial y^2}Y(y) = -\frac{2mE_2}{\hbar^2}Y(y). \tag{10.32}$$

The solutions of these equations are given by

$$X(x) = \frac{\sqrt{2}}{\sqrt{a}}\sin\left(\frac{n_x\pi}{a}x\right),\ E_1 = \frac{n_x^2\pi^2\hbar^2}{2ma^2}, \tag{10.33}$$

$$Y(y) = \frac{\sqrt{2}}{\sqrt{a}}\sin\left(\frac{n_y\pi}{a}y\right),\ E_y = \frac{n_y^2\pi^2\hbar^2}{2mb^2}. \tag{10.34}$$

Finally, the solution becomes

$$\psi(x, y) = \frac{2}{\sqrt{a}\sqrt{b}}\sin\left(\frac{n_x\pi}{a}x\right)\sin\left(\frac{n_y\pi}{b}y\right), \tag{10.35}$$

and the energy eigenvalues are defined as

$$E = \frac{\pi^2\hbar^2}{2m}\left(\frac{n_x^2}{a^2} + \frac{n_y^2}{b^2}\right). \tag{10.36}$$

These results are important in that the ground state corresponds to the situation of $n_x = 1$ and $n_y = 1$. The wave function and the energy eigenvalues are given by

$$\psi(x, y) = \frac{2}{\sqrt{a}\sqrt{b}}\sin\left(\frac{\pi}{a}x\right)\sin\left(\frac{\pi}{b}y\right), \tag{10.36}$$

$$E = \frac{\pi^2\hbar^2}{2m}\left(\frac{1}{a^2} + \frac{1}{b^2}\right). \tag{10.38}$$

Now, if we take $a = b$ and we take into account $n_x = 1$, $n_y = 2$ and $n_x = 2$, $n_y = 1$, then we see that there is a degenerate state for this situation. The degenerate situation can be briefly summarised as follows. If a quantum state has the same energy eigenvalues but different wave functions, then it is a degenerate state. This problem is a good example to analyse these degenerate states. Now we can analyse the above cases in Julia language code. This analysis is done in two different ways. In the first case, a step-by-step and somewhat lengthy method is described. The second application is relatively short and more compact in defining the boundary conditions.

10.6.1 The First Way

We start with the importing *SymPy* module, variables and the functions (Programmes 10.11, 10.12, 10.13, 10.14 and 10.15).

The wave function X(x) is normalised by the above procedures and determined together with the energy eigenvalues. The same steps are given below for Y(y) with appropriate variable names (Programmes 10.16 and 10.17).

Now we can write $\psi(x, y)$ wave function as following. Moreover, we can write different quantum states. First three states are searched as following with its plots (Programmes 10.18, 10.19, 10.20 and Fig. 10.3).

```
[1]: using SymPy

[2]: x = symbols(:x)
     y = symbols(:y)
     h = symbols(:h, real = true, positive = true)
     m = symbols(:m, real = true, positive = true)
     E = symbols(:E, real = true, positive = true)
     a = symbols(:a, real = true, positive = true)
     b = symbols(:b, real = true, positive = true)

     E1 = symbols(:E1, real = true, positive = true)
     E2 = symbols(:E2, real = true, positive = true)

     X = SymFunction("X")(x)
     Y = SymFunction("Y")(y)

     Ψ = X(x) * Y(y)
```

[2]: $X(x)Y(y)$

```
[3]: F = Eq((-h^2/(2m)) * (diff(Ψ, x, x) + diff(Ψ, y, y)), E * Ψ)
```

[3]: $-\frac{h^2\left(X(x)\frac{d^2}{dy^2}Y(y)+Y(y)\frac{d^2}{dx^2}X(x)\right)}{2m} = EX(x)Y(y)$

Programme 10.11 The importing *SymPy*, variables and functions

```
[7]: fx = Eq((-h^2/(2m)) * (diff(Ψ, x, x) / (X * Y)), E1)
```

[7]: $-\frac{h^2\frac{d^2}{dx^2}X(x)}{2mX(x)} = E_1$

```
[8]: sol_fx = dsolve(fx)
```

[8]: $X(x) = C_1\sin\left(\frac{\sqrt{2}\sqrt{E_1}\sqrt{m}x}{h}\right) + C_2\cos\left(\frac{\sqrt{2}\sqrt{E_1}\sqrt{m}x}{h}\right)$

```
[9]: sol_fx2 = sol_fx.rhs()
```

[9]: $C_1\sin\left(\frac{\sqrt{2}\sqrt{E_1}\sqrt{m}x}{h}\right) + C_2\cos\left(\frac{\sqrt{2}\sqrt{E_1}\sqrt{m}x}{h}\right)$

Programme 10.12 The Schrödinger equation for X(x) and its solution

```
[10]: c2 = free_symbols(sol_fx2)[2]
```

[10]: C_2

```
[11]: c2v = solve(sol_fx2(x => 0) ~ 0, c2)[1]
```

[11]: 0

```
[12]: fx1 = sol_fx2(c2 => Sym(c2v[1]))
```

[12]: $C_1 \sin\left(\frac{\sqrt{2}\sqrt{E_1}\sqrt{m}x}{h}\right)$

```
[13]: fx2 = simplify(integrate(fx1 * fx1, x))
```

[13]: $\frac{C_1^2 x}{2} - \frac{\sqrt{2}C_1^2 h \sin\left(\frac{2\sqrt{2}\sqrt{E_1}\sqrt{m}x}{h}\right)}{8\sqrt{E_1}\sqrt{m}}$

```
[14]: fxn = subs(fx2, (h, 0))
```

[14]: $\frac{C_1^2 x}{2}$

Programme 10.13 Determination of the constants

```
[15]: fx3 = Eq(subs(fxn, (x, a)), 1)
```

[15]: $\frac{C_1^2 a}{2} = 1$

```
[16]: c1 = free_symbols(fx3)[1]
      solve(fx3, c1)

      c1v = solve(fx3, c1)
```

[16]: $\begin{bmatrix} -\frac{\sqrt{2}}{\sqrt{a}} \\ \frac{\sqrt{2}}{\sqrt{a}} \end{bmatrix}$

```
[17]: fxh = fx1(c1 => Sym(c1v[2]))
```

[17]: $\frac{\sqrt{2}\sin\left(\frac{\sqrt{2}\sqrt{E_1}\sqrt{m}x}{h}\right)}{\sqrt{a}}$

Programme 10.14 Determination of the wave function for X(x)

```
[19]: fsol = solve(subs(fxh, x => a))

      n = symbols(:n, integer = true)
      p = symbols(:p) # equals to pi

      # vec = [Dict(:E1 => n^2*p^2*h^2/(2*a^2*m))]
      E1_f = n^2 * fsol[1][E1]
```

[19]: $\frac{\pi^2 h^2 n^2}{2a^2 m}$

Programme 10.15 The energy eigenvalues for X(x)

10.6.2 The Second Way

See Programmes 10.21, 10.22, 10.23, 10.24 and 10.25.

```
[20]: fy = Eq((-h^2/(2m)) * (diff(Ψ, y, y) / (X * Y)), E2)
```

[20]: $$-\frac{h^2 \frac{d^2}{dy^2} Y(y)}{2mY(y)} = E_2$$

```
[21]: sol_fy = dsolve(fy)

      sol_fy2 = sol_fy.rhs()
```

[21]: $$C_1 \sin\left(\frac{\sqrt{2}\sqrt{E_2}\sqrt{m}y}{h}\right) + C_2 \cos\left(\frac{\sqrt{2}\sqrt{E_2}\sqrt{m}y}{h}\right)$$

```
[22]: c2 = free_symbols(sol_fx2)[2]

      c2v = solve(sol_fx2(x => 0) ~ 0, c2)[1]

      fy1 = sol_fy2(c2 => Sym(c2v[1]))
```

[22]: $$C_1 \sin\left(\frac{\sqrt{2}\sqrt{E_2}\sqrt{m}y}{h}\right)$$

```
[23]: fyn = simplify(integrate(fy1*fy1, y))
```

[23]: $$\frac{C_1^2 y}{2} - \frac{\sqrt{2} C_1^2 h \sin\left(\frac{2\sqrt{2}\sqrt{E_2}\sqrt{m}y}{h}\right)}{8\sqrt{E_2}\sqrt{m}}$$

Programme 10.16 The Schrödinger equation for Y(y) and its solution

```
[25]: fy1 = sol_fy2(c1 => Sym(c1v[2]))

      c2 = free_symbols(fy1)[1]

      fy1 = subs(fy1, c2 => 0)

      fsol2 = solve(subs(fy1, y => b))

      l = symbols(:l, integer = true)
      p = symbols(:p) # equals to pi

      # vec = [Dict(:E1 => n^2*p^2*h^2/(2*a^2*m))]
      E2_f = l^2 * fsol2[1][E2]
```

[25]: $$\frac{\pi^2 h^2 l^2}{2b^2 m}$$

```
[26]: fy1 = subs(fy1, c2 => 0)
```

[26]: $$\frac{\sqrt{2}\sin\left(\frac{\sqrt{2}\sqrt{E_2}\sqrt{m}y}{h}\right)}{\sqrt{b}}$$

```
[27]: fxh
```

[27]: $$\frac{\sqrt{2}\sin\left(\frac{\sqrt{2}\sqrt{E_1}\sqrt{m}x}{h}\right)}{\sqrt{a}}$$

```
[28]: fy1
```

[28]: $$\frac{\sqrt{2}\sin\left(\frac{\sqrt{2}\sqrt{E_2}\sqrt{m}y}{h}\right)}{\sqrt{b}}$$

Programme 10.17 The solution for Y(y)

```
[30]: Ψ = subs(Ψ, Dict((X(x) => fxh), (Y(y) => fy1)))
```

[30]: $$\frac{2\sin\left(\frac{\sqrt{2}\sqrt{E_1}\sqrt{m}x}{\hbar}\right)\sin\left(\frac{\sqrt{2}\sqrt{E_2}\sqrt{m}y}{\hbar}\right)}{\sqrt{a}\sqrt{b}}$$

```
[31]: Ψ = expand(subs(Ψ, Dict((E1 => E1_f), (E2 => E2_f))))
```

[31]: $$\frac{2\sin\left(\frac{\pi x|n|}{a}\right)\sin\left(\frac{\pi y|l|}{b}\right)}{\sqrt{a}\sqrt{b}}$$

Programme 10.18 The general solution for the problem

```
[32]: Ψ = expand(subs(Ψ, Dict((a => 1), (b => 1))))
```

[32]: $2\sin(\pi x|n|)\sin(\pi y|l|)$

```
[33]: Ψ11 = expand(subs(Ψ, Dict((n => 1), (l => 1))))
```

[33]: $2\sin(\pi x)\sin(\pi y)$

```
[34]: Ψ12 = expand(subs(Ψ, Dict((n => 1), (l => 2))))
```

[34]: $2\sin(\pi x)\sin(2\pi y)$

```
[35]: Ψ21 = expand(subs(Ψ, Dict((n => 2), (l => 1))))
```

[35]: $2\sin(2\pi x)\sin(\pi y)$

Programme 10.19 Different quantum states in two-dimensional potential well

```
[36]: using Plots

x_vals = LinRange(0, 1, 100)
y_vals = LinRange(0, 1, 100)

z_vals0 = [Ψ11(x, y) for x in x_vals, y in y_vals]
z_vals1 = [Ψ21(x, y) for x in x_vals, y in y_vals]
z_vals2 = [Ψ12(x, y) for x in x_vals, y in y_vals]

surface(x_vals, y_vals, z_vals0, xlabel="x", ylabel="y", zlabel="Ψ11(x,y) : $Ψ11",
    title="Ground state", size=(500, 500))
```

[36]:

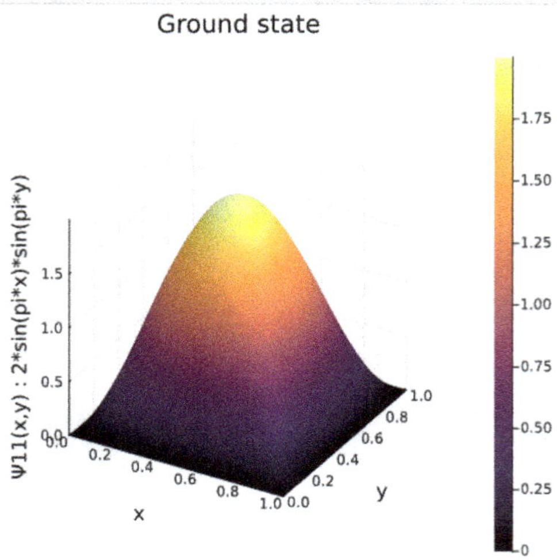

Programme 10.20 The ground state of the system

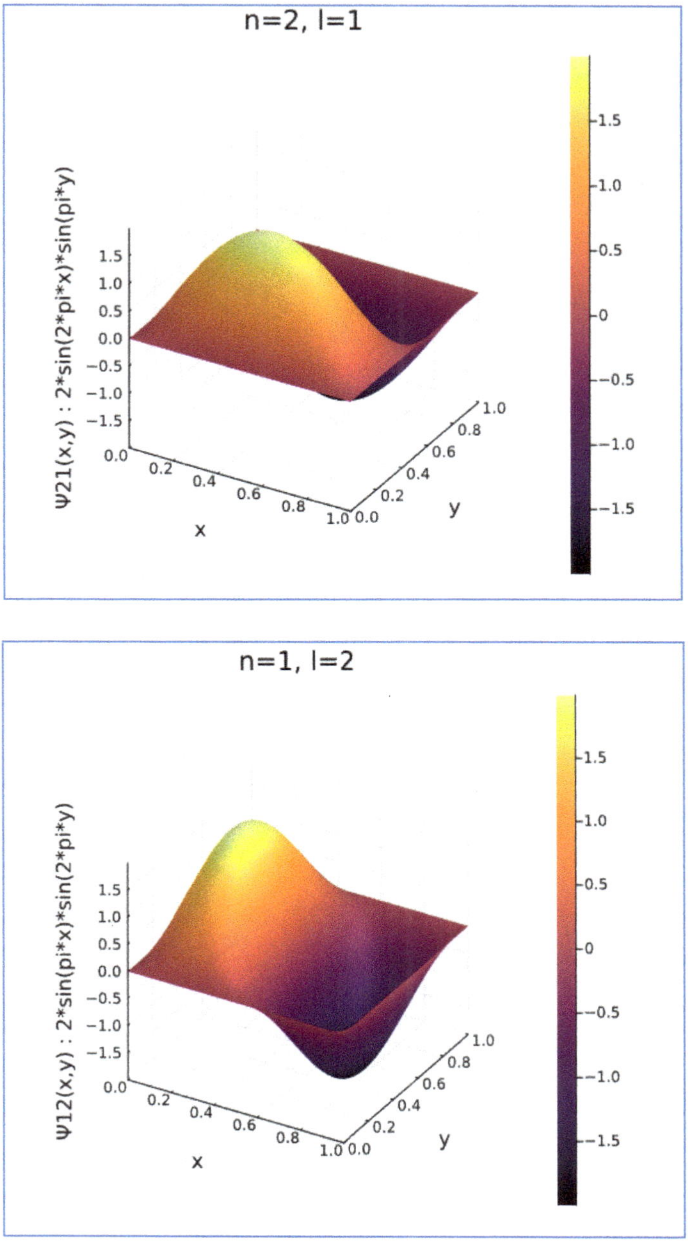

Fig. 10.3 $n = 2, l = 1$ and $n = 1, l = 2$ states (degenerate states)

[10]:
```
fx = Eq((-h^2/(2m)) * (diff(Ψ, x, x) / (X * Y)), E1)
```

[10]: $-\frac{\hbar^2 \frac{d^2}{dx^2} X(x)}{2m X(x)} = E_1$

[14]:
```
solution_X = dsolve(fx, X)

X_general = solution_X.rhs

boundary_condition_1 = Eq(X_general.subs(x, 0), 0)
boundary_condition_2 = Eq(X_general.subs(x, a), 0)

fx_sol = solve([boundary_condition_1, boundary_condition_2])

n = symbols(:n, integer = true)
p = symbols(:p) # equals to pi

n = symbols(:n, integer = true)
p = symbols(:p) # equals to pi

E1_f = n^2 * fx_sol[2][E1]
```

[14]: $\frac{\pi^2 \hbar^2 n^2}{2a^2 m}$

Programme 10.21 The equation for x after importing *SymPy* and the variables

[16]:
```
X_general

c1 = free_symbols(X_general)[1]
c2 = free_symbols(X_general)[2]

X = subs(X_general, Dict((c2 => 0), (c1 => sqrt(2/a))))
```

[16]: $\frac{\sqrt{2}\sin\left(\frac{\sqrt{2}\sqrt{E_1}\sqrt{m}x}{\hbar}\right)}{\sqrt{a}}$

Programme 10.22 The wave function for X(x)

[18]:
```
fy = Eq((-h^2/(2m)) * (diff(Y, y, y)), E2*Y)
```

[18]: $-\frac{\hbar^2 \frac{d^2}{dy^2} Y(y)}{2m} = E_2 Y(y)$

[20]:
```
solution_Y = dsolve(fy, Y)

Y_general = solution_Y.rhs

boundary_condition_1 = Eq(Y_general.subs(y, 0), 0)
boundary_condition_2 = Eq(Y_general.subs(y, b), 0)

fy_sol = solve([boundary_condition_1, boundary_condition_2])

l = symbols(:l, integer = true)
p = symbols(:p) # equals to pi

# vec = [Dict(:E1 => n^2*p^2*h^2/(2*a^2*m))]
E2_f = l^2 * fy_sol[2][E2]
```

[20]: $\frac{\pi^2 \hbar^2 l^2}{2b^2 m}$

[22]:
```
Y_general

c1 = free_symbols(Y_general)[1]
c2 = free_symbols(Y_general)[2]

Y = subs(Y_general, Dict((c2 => 0), (c1 => sqrt(2/b))))
```

[22]: $\frac{\sqrt{2}\sin\left(\frac{\sqrt{2}\sqrt{E_2}\sqrt{m}y}{\hbar}\right)}{\sqrt{b}}$

Programme 10.23 The equation and its solution for Y(y)

```
[24]: Ψ = X * Y
```

$$[24]:\ \frac{2\sin\left(\frac{\sqrt{2}\sqrt{E_1}\sqrt{m}x}{\hbar}\right)\sin\left(\frac{\sqrt{2}\sqrt{E_2}\sqrt{m}y}{\hbar}\right)}{\sqrt{a}\sqrt{b}}$$

```
[26]: E = simplify(E1_f + E2_f)
```

$$[26]:\ \frac{\pi^2\hbar^2(a^2l^2+b^2n^2)}{2a^2b^2m}$$

```
[28]: Ψ = expand(subs(Ψ, Dict((E1 => E1_f), (E2 => E2_f))))
```

$$[28]:\ \frac{2\sin\left(\frac{\pi x|n|}{a}\right)\sin\left(\frac{\pi y|l|}{b}\right)}{\sqrt{a}\sqrt{b}}$$

Programme 10.24 The general solution and the energy eigenvalues

```
[30]: Ψ = expand(subs(Ψ, Dict((a => 1), (b => 4))))
```

$$[30]:\ \sin(\pi x|n|)\sin\left(\frac{\pi y|l|}{4}\right)$$

```
[32]: Ψ11 = expand(subs(Ψ, Dict((n => 1), (l => 1))))
```

$$[32]:\ \sin(\pi x)\sin\left(\frac{\pi y}{4}\right)$$

```
[34]: Ψ12 = expand(subs(Ψ, Dict((n => 1), (l => 2))))
```

$$[34]:\ \sin(\pi x)\sin\left(\frac{\pi y}{2}\right)$$

```
[36]: Ψ21 = expand(subs(Ψ, Dict((n => 2), (l => 1))))
```

$$[36]:\ \sin(2\pi x)\sin\left(\frac{\pi y}{4}\right)$$

Programme 10.25 The quantum states for $a = 1$ and $b = 4$

10.7 Problems

1. Determine the mathematical equivalent of the wave packet that will be formed in a case where 4 waves with a frequency difference of 0.1 units are superimposed and try to perform this operation for Julia codes and trigonometric transformations. What can you say about such a wave packet.
2. Write the Schrödinger equation for a free particle in free space and try to solve this equation using Julia codes. Try to calculate the expected value of momentum and energy for a free particle using Julia codes.
3. The potential barrier was analysed above. Determine the wave function in each region and try to calculate the probability of finding a particle in these regions. Note that the sum of these probabilities must be a unit value.
4. The probability of the particle passing into the third region in the case of a potential barrier was calculated. Calculate the probability of reflection of the particle to the first region using Julia codes. Verify that the sum of these probabilities must be 1.

5. Try to calculate the expected value of the momentum and energy of the particle in all three regions for the potential barrier with the codes you will write in Julia language.
6. Calculate the expected states of the momentum and energy of the particle for the first 4 quantum states for a one-dimensional potential well. For these first 4 quantum states, use Julia codes to calculate the most probable locations of the particle and discuss your results by integrating the square of the wave function for these quantum states.

References

1. Kenneth S. Krane. 2019. Modern Physics, 4th Edition. Wiley. ISBN: 978-1-119-49546-8.
2. Julia Documentation. https://julialang.org/
3. SymPy module. https://docs.sympy.org/latest/tutorials/intro-tutorial/index.html
4. Plots module. https://docs.juliaplots.org/stable/tutorial/#tutorial
5. Mehmet Zengin, R. Gökhan Türeci. 2008. Kuantum Fiziği Problemleri ve Çözümleri. Bilim Yayıncılık, Ankara. ISBN: 978-975556-065-6.

Chapter 11
Julia Programming Applications in Nuclear Physics

11.1 Introduction

The experiments of Rutherford and his colleagues, who investigated the scattering of alpha particles from a thin gold plate, were the most important experiments that led to the beginning of nuclear physics. In the process, after the discovery of the atomic nucleus, models were developed that the atomic nucleus consists of protons and electrons. However, according to the Heisenberg uncertainty principle as the most practical way, if there were electrons in the nucleus, they should come out of the nucleus with energies on the order of MeV, but such a result could not be reached. Eventually Chadwick discovered the neutron, which has a mass slightly larger than the mass of the proton and has no charge. The fact that the neutron is uncharged and can interact with the atomic nucleus without being affected by the Coulomb barrier of the nucleus has paved the way for the nuclear technology we use today, as well as for many other scientific research. In this chapter, a simple overview of Nuclear Physics, elegantly covered by authors such as Krane [1] and Lilley [2], will be explored using the Julia programming language.

11.2 Relationship Between Proton Number and Mass Number for Light Nuclei

We know that Z protons and N protons are formed in the atomic nucleus and that the sum of the number of protons and neutrons gives the mass number of the nucleus, A. As the atomic nucleus grows, that is, as the number of protons increases, an excess of neutrons is formed for large nuclei in order to balance the Coulomb repulsion between protons.

On the other hand, the energy that must be expended to bring an atomic nucleus together with protons and neutrons is smaller than the total energy of the particles in

R. G. Türeci et al., *Julia Programming for Physics Applications*,
https://doi.org/10.1007/978-3-031-84716-5_11

```
[1]: using SymPy

[2]: z = symbols(:z, integer = true, positive = true)
     n = symbols(:n, integer = true, positive = true)
     A = symbols(:A, integer = true, positive = true)

     a_v = symbols(:a_v)
     a_s = symbols(:a_s)
     a_c = symbols(:a_c)
     a_sym = symbols(:a_sym)
     a_p = symbols(:a_p)

     B = SymFunction("B")

[2]: B
```

Programme 11.1 The importing of the module and the variable definitions

the free state, and the difference is defined as the binding energy. While it is relatively easy to calculate the binding energy for gravitational force or electromagnetic interacting particles, it is not possible to calculate this energy for atomic nuclei since all the properties of the nuclear force are not yet fully known. For this reason, the semi-empirical binding energy, written from experimental observations and theoretical considerations, is widely used in nuclear physics.

The semi-empirical binding energy is given by

$$B = a_V A - a_S A^{2/3} - a_C Z^2 A^{-1/3} - a_{sim}(Z - N)^2 A^{-1} + \delta a_P A^{-3/4}. \tag{11.1}$$

Using Eq. (11.1), let us examine that for light atomic nuclei the number of protons should be around half the mass number.

Clearly, we need to find the maximum value of the semi-empirical binding energy with respect to the number of protons and set it equal to zero. We want to do this using the Julia code. First two cells in Programme 11.1 are used to import the SymPy module and to define the variables (Programme 11.2).

This solution shows that the atomic number equals to the neutron number for the small mass number.

11.3 The Coulomb Energy of a Uniform Spherical Nuclei

If we consider the atomic nucleus as a uniform sphere and assume that the protons are uniformly distributed in this sphere, we can calculate the Coulomb energy of the nucleus. Programme 11.3 gives the Julia code needed to calculate this energy by integrating symbolically.

```
[3]: B = a_v*A - a_s*A^(2//3) - a_c*z^2/A^(1//3) - a_sym*(z-n)^2/A + a_p*A^(-1//2)
```

[3]: $-A^{\frac{2}{3}}a_s + Aa_v - \frac{a_{sym}(-n+z)^2}{A} + \frac{a_p}{\sqrt{A}} - \frac{a_c z^2}{\sqrt[3]{A}}$

```
[4]: dB = diff(B, z)
```

[4]: $-\frac{a_{sym}(-2n+2z)}{A} - \frac{2a_c z}{\sqrt[3]{A}}$

```
[5]: f = Eq(dB, 0)
```

[5]: $-\frac{a_{sym}(-2n+2z)}{A} - \frac{2a_c z}{\sqrt[3]{A}} = 0$

```
[6]: f_sol = solve(f, z)
```

[6]: $\left[\frac{\sqrt[3]{A}a_{sym}n}{\sqrt[3]{A}a_{sym} + Aa_c}\right]$

```
[7]: fx = expand(f_sol[1])
```

[7]: $\frac{\sqrt[3]{A}a_{sym}n}{\sqrt[3]{A}a_{sym} + Aa_c}$

```
[8]: fxx = factor(subs(fx, Dict((a_c => 0.687), (a_sym => 22.409))))
```

[8]: $\frac{1.0\sqrt[3]{A}n}{1.0\sqrt[3]{A} + 0.0306573251818466A}$

Programme 11.2 The solution of the derivation of the binding energy according to the atomic number

```
[1]: using SymPy

[2]: z = symbols(:z)
     ρ = symbols(:ρ)
     p = symbols(:π)
     e = symbols(:e)
     R = symbols(:R)
     r = symbols(:r)
```

[2]: r

```
[3]: fe = (4//3)*p*r^3*ρ * 4p*r^2*ρ/r
```

[3]: $\frac{16r^4\pi^2\rho^2}{3}$

```
[4]: E = integrate(fe, (r, 0, R))
```

[4]: $\frac{16R^5\pi^2\rho^2}{15}$

```
[5]: E = subs(E, ρ => (3//4)*(z * e)/(p*R^3))
```

[5]: $\frac{3e^2z^2}{5R}$

```
[6]: E_c = symbols(:E_c)
```

[6]: E_c

```
[7]: E_c = simplify(E - E/z)
```

[7]: $\frac{3e^2z(z-1)}{5R}$

```
[8]: E_c |> string

[8]: "3*e^2*z*(z - 1)/(5*R)"
```

Programme 11.3 The Coulomb energy a uniform nucleus

11.4 The Average Decay Time of a Nucleus Undergoing Two Different Decays

If the radioactive decay of a nucleus takes place with two different decay rates and decay constants λ_1 and λ_2, the number of nuclei in a nucleus is given by

$$N(t) = N_0 \exp(-(\lambda_1 + \lambda_2))t. \tag{11.2}$$

Let us try to find the mean life expression of the decay of a nucleus exhibiting such a decay. We can write the following expressions for the first and second type decays.

$$dN_1 = -\lambda_1 N dt, \tag{11.3}$$

$$dN_2 = -\lambda_2 N dt. \tag{11.4}$$

Thus, the following expression can be written for the change of the nucleus per unit time.

$$dN = -(\lambda_1 + \lambda_2) N dt. \tag{11.5}$$

Programme 11.4 gives the definitions, the differential equation and its solution. The looking for result is given in Programme 11.5.

```
[1]: using SymPy

[2]: t = symbols(:t)

     λ_1 = symbols(:λ_1)
     λ_2 = symbols(:λ_2)

     n = SymFunction("n")(t)

[2]: n(t)

[3]: f = Eq(diff(n, t), -n*(λ_1 + λ_2))
```

[3]: $\frac{d}{dt}n(t) = -(\lambda_1 + \lambda_2)n(t)$

```
[4]: sol = dsolve(f)
```

[4]: $n(t) = C_1 e^{-t(\lambda_1+\lambda_2)}$

```
[5]: f = sol.rhs()
```

[5]: $C_1 e^{-t(\lambda_1+\lambda_2)}$

Programme 11.4 The related definitions and the solution of the differential equation

```
[6]: c1 = free_symbols(f)[1]
```

[6]: C_1

```
[7]: n_0 = symbols(:n_0)
     c1v = solve(f(t => 0) ~ n_0)[1]
```

[7]: $\{C_1 \quad => \quad n_0$

```
[8]: c1_v = c1v[c1]
```

[8]: n_0

```
[9]: n = f(c1 => c1v[c1])
```

[9]: $n_0 e^{-t(\lambda_1+\lambda_2)}$

```
[10]: τ = symbols(:τ)
      f1 = subs(n, t => τ)
```

[10]: $n_0 e^{-\tau(\lambda_1+\lambda_2)}$

```
[11]: qq = Eq(f1, c1v[c1]/exp(1))
```

[11]: $n_0 e^{-\tau(\lambda_1+\lambda_2)} = 0.367879441171442 n_0$

```
[12]: solve(qq, τ)[1]
```

[12]: $\frac{1.0}{\lambda_1+\lambda_2}$

Programme 11.5 The related definitions and the solution of the differential equation

11.5 A Radioactive Nucleus Bombarded with Neutrons

Let us assume that a radioactive nuclide with decay constant λ is bombarded with neutrons and produced at the rate of formation with $N\sigma\phi$ where N corresponds to the target nucleus number, σ corresponds to the cross section which is dependent to the neutron energies in unit of cm^2 and ϕ corresponds to the neutron flux. We can say that this radioactive nuclide is bombarded with neutrons on the one hand and decays with the constant lam on the other.

After being bombarded with neutrons, the number of radioactive nuclides at the end of time t is given by

$$N_2 = \frac{N\sigma\phi}{\lambda}(1 - \exp(-\lambda t)). \tag{11.6}$$

We want to derive this relation. Radioactive nuclei are produced and decay at the same time. We must therefore take these into account in the change of the nuclei produced over time. Then we can write the following equation.

$$dN_2 = N\sigma\phi dt - \lambda N_2 dt. \tag{11.7}$$

We want to solve Eq. (11.7) using Julia codes. Again, let's start with importing the SymPy module and defining variables.

The differential equation is defined in the third cell in Programme 11.6. The solution of this equation is given in 4th cell in the programme. Now we should apply the boundary conditions. Thus, we can find the looking for solution (Programme 11.7).

```
[1]: using SymPy

[2]: n = symbols(:n, real = true)
     σ = symbols(:σ, real = true)
     Φ = symbols(:Φ, real = true)
     λ = symbols(:λ, real = true)
     t = symbols(:t, real = true)

     n2 = SymFunction("n2")(t)
```

[2]: $n_2(t)$

```
[3]: f = Eq(diff(n2, t), -λ*n2 + n*σ*Φ)
```

[3]: $\frac{d}{dt}n_2(t) = n\Phi\sigma - \lambda n_2(t)$

```
[4]: sol = dsolve(f, n2)
```

[4]: $n_2(t) = C_1e^{-t\lambda} + \frac{n\Phi\sigma}{\lambda}$

```
[5]: f = sol.rhs()
```

[5]: $C_1e^{-t\lambda} + \frac{n\Phi\sigma}{\lambda}$

```
[6]: c1 = free_symbols(f)[1]
```

[6]: C_1

```
[7]: n_0 = symbols(:n_0)

     c1v = solve(f(t => 0) ~ 0)[1]
```

[7]: $\{C_1 \;=>\; -\frac{n\Phi\sigma}{\lambda}$

Programme 11.6 The definitions and the solving of the differential equation

```
[8]: c1_v = c1v[c1]
```

[8]: $-\frac{n\Phi\sigma}{\lambda}$

```
[9]: n2 = f(c1 => c1v[c1])
```

[9]: $\frac{n\Phi\sigma}{\lambda} - \frac{n\Phi\sigma e^{-t\lambda}}{\lambda}$

```
[10]: n2
```

[10]: $\frac{n\Phi\sigma}{\lambda} - \frac{n\Phi\sigma e^{-t\lambda}}{\lambda}$

Programme 11.7 The looking for solution

11.6 Problems

1. Write the Julia programme to calculate the neutron separation energy of a $^4He_2$ nucleus and a $^7Li_3$ nucleus. How can you explain the results from a physical point of view?
2. Solve the first question for $^{16}O_8$ and $^{20}O_8$, but this time for the proton splitting energy and compare the results physically.
3. Define 1 electron volt in Julia and express the energy of an electron with an energy of 20 MeV leaving the nucleus in terms of erg and Joule. *Ops!* Can electrons be found in the nucleus? Can you answer this question using the Heisenberg uncertainty principle and write a simple Julia code for it?
4. Solve the differential equation defining the radioactive decay law for the initial condition of N_0 nuclei at time $t = 0$.
5. Assume that in a mass spectrometer particles with masses m_1, m_2 and m_3 are collected in three different compartments with orbits of radii r, $3/2r$ and $4r$. Write a Julia program that specifies the relationship between the masses m_1, m_2 and m_3 for the if conditional. Assume that the kinetic energy of the particles is the same.

References

1. Kenneth S. Krane. 1991. Introductory Nuclear Physics, 3rd Edition, Wiley. ISBN: 978-0-471-80553-3.
2. John Lilley. 2013. Nuclear Physics: Principles and Applications, Wiley. ISBN: 978-0-471-97936-4.

Chapter 12
Julia Programming Applications in Solid State Physics

12.1 Introduction

In this section, we will discuss a few of the basic concepts of solid state physics and try to analyse the calculations related to these concepts with the Julia programming language. Firstly, we will explain the physical basis of the concepts, then we will move on to programme writing and some applications will be mentioned.

12.2 Wigner Seitz Unit Cell

The Brillouin zone is an important concept in understanding the energy band structures of crystal structures and in analysing their electronic properties. The Brillouin zone is a specific region of momentum space in a crystalline structure and is usually defined depending on the symmetry of a crystal.

Since the atoms of a crystal are arranged in an orderly manner, the electrons follow a certain symmetrical arrangement based on this structure. Electrons move in momentum space (or k-space) and the boundaries of this space do not extend beyond the Brillouin zone.

The most widely used diffraction condition in solid state physics is expressed by Brillouin, which is very useful for the band theory of electrons and other elementary applications.

$$2\overrightarrow{k} \cdot \overrightarrow{G} = G^2 \tag{12.1}$$

The parallelogram prism defined by the axes a_1, a_2, a_3 is called a primitive cell. The primitive unit cell fills the entire space by repeating the crystal translation process.

With u_1, u_2, u_3 being integers, the smallest cell that can be crystallised is formed by the primitive translation vectors $\vec{a}_1, \vec{a}_2$ and $\vec{a}_3$. The lattice translation process is

R. G. Türeci et al., *Julia Programming for Physics Applications*,
https://doi.org/10.1007/978-3-031-84716-5_12

defined by the crystal translation vector denoted in Eq. 11.3.

$$\vec{T} = u_1 \vec{a}_1 + u_2 \vec{a}_2 + u_3 \vec{a}_3 \tag{12.2}$$

A Brillouin zone is defined as a Wigner-Seitz primitive cell in the inverse lattice. The inverse lattice vectors of a structure defined by the axes a_1, a_2, a_3 are obtained by the Eq. 12.3.

$$\begin{aligned}\vec{b}_1 &= 2\pi \frac{\vec{a}_2 \times \vec{a}_3}{\vec{a}_1 \cdot \vec{a}_2 \times \vec{a}_3} \\ \vec{b}_2 &= 2\pi \frac{\vec{a}_3 \times \vec{a}_1}{\vec{a}_1 \cdot \vec{a}_2 \times \vec{a}_3} \\ \vec{b}_3 &= 2\pi \frac{\vec{a}_1 \times \vec{a}_2}{\vec{a}_1 \cdot \vec{a}_2 \times \vec{a}_3}\end{aligned} \tag{12.3}$$

The most general translation vector for the inverse lattice is given by the Eq. 12.4. v_1, v_2, v_3 in this expression are integers and the $\vec{G}$ is called the inverse lattice vector. The inverse lattice is the lattice of the crystal in Fourier space.

The dimension of the vectors in the real lattice is the $\left[\text{length}\right]$, and the dimension of the vectors in the inverse lattice is of order $\left[1/\text{length}\right]$.

$$\vec{G} = v_1 \vec{b}_1 + v_2 \vec{b}_2 + v_3 \vec{b}_3 \tag{12.4}$$

This example shows how to create and visualise a Wigner-Seitz cell for a simple cubic crystal structure [1, 2]. For more complex crystal structures, consider using libraries such as ***CrystalViewer.jl***.

CrystalViewer.jl is a package developed in the Julia programming language for visualising and analysing crystal structures. This package is particularly useful for researchers and engineers working with crystal structures because it allows to understand the geometry and symmetry of a crystal. It also allows visual presentation of various physical and chemical properties of crystals [3, 4] (Programmes 12.1, 12.2 and 12.3).

12.3 Bragg Law

Bragg's Law is a fundamental principle in crystallography and is used to understand and describe how X-rays or other types of electromagnetic radiation interact with crystalline materials. It explains the conditions under which constructive interference occurs when X-rays are scattered by the regularly spaced planes of atoms in a crystal. This principle is essential in determining the structure of materials using X-ray diffraction.

```
[157]: using Plots

       println("Enter lattice parameter : ")
       a = parse(Float64, readline())

       Enter lattice parameter :
       stdin>  2

[157]: 2.0

[159]: println("Enter number of points) : ")
       num_p = parse(Float64, readline())

       Enter number of points) :
       stdin>  12

[159]: 12.0
```

Programme 12.1 Definition of variables

```
[161]:

       function hexa_lat(a, num_p)
           points = []
           for i in 0:num_p-1
               for j in 0:num_p-1
                   x = a * (i + 0.5 * (j % 2))  # x coordinates for hexagonal lattice
                   y = a * (j * sqrt(3) / 2)    # y coordinates for hexagonal lattice
                   push!(points, (x, y))
               end
           end
           return points
       end

       function plot_lattice(points, title)
           xs = [p[1] for p in points]
           ys = [p[2] for p in points]
           scatter(xs, ys, label="Lattice points", legend=:topright, title=title)
           xlims!(-2, 10)
           ylims!(-2, 10)
           xlabel!("X axis")
           ylabel!("Y axis")
       end

       # Determination of Hexagonal Lattice
       hexagonal_lattice_points = hexa_lat(a, num_p)

       # Lattice
       plot_lattice(hexagonal_lattice_points, "Hexagonal Lattice")
```

Programme 12.2 Collection data and plot part of programme

The crystal structure is analysed by diffraction of photons, neutrons and electrons. Diffraction depends on the wavelength and crystal structure. At optical wavelengths up to 5000 A, optical diffraction occurs as a result of the scattering of a wave from each of the crystal atoms. If the wavelength of the incident radiation is the size of the lattice constant or smaller, diffraction can occur in very different directions.

If we assume that the parallel lattice planes are bounded by intervals d, the path difference between the rays reflected from two neighbouring planes will be $2d\sin\theta$,

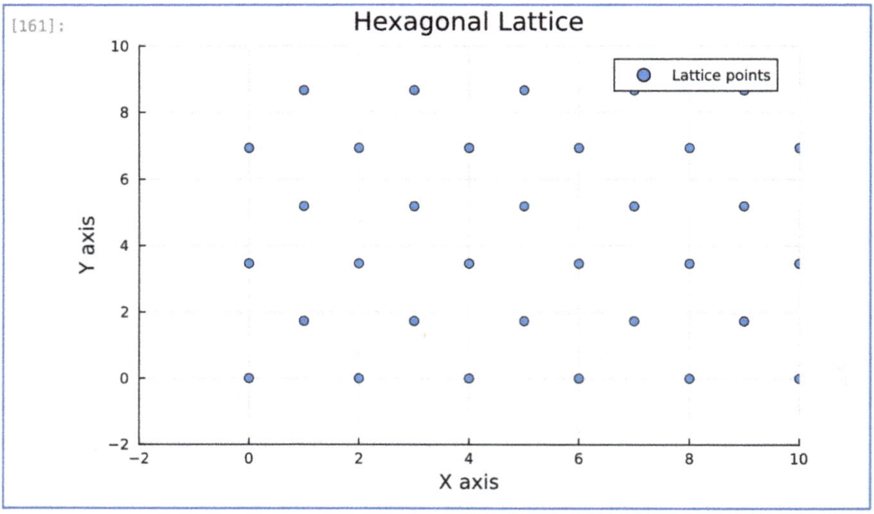

Programme 12.3 Output for hexagonal lattice

where θ is the angle with the plane. For constructive interference to occur, these path differences between the rays reflected from successive planes must be integer multiples of the wavelength.

$$2d\sin\theta = n\lambda \tag{12.5}$$

For brag's law to be realised, $\lambda \leq 2d$ must be. Therefore, visible light is not used [1, 2].

Programme 12.4 symbolises the variables and function for bragg angle by using funtions [3, 4] (Programme 12.5).

12.4 Hall Effect

If we consider motion in a uniform magnetic flux, the Lorentz force acting on an electron is given by Eq. 12.6.

$$\overrightarrow{F} = -(\overrightarrow{E} + \frac{1}{c}(\overrightarrow{v} \times \overrightarrow{B})) \tag{12.6}$$

If we take $m\overrightarrow{v} = \hbar\delta\overrightarrow{k}$ is Eq. 12.6, the equation of motion $\hbar(\frac{d}{dt} + \frac{1}{\tau})\delta\overrightarrow{k} = \overrightarrow{F}$,will return to Eq. 12.7.

```
[23]: using Printf
      using LinearAlgebra

      wavelength = 1.5e-11  # (in meters)

      println("Interplanar spacing (in meters): ")
      d = parse(Float64, readline())

      println("Maximum diffraction number : ")
      max = parse(Int, readline())


      # Function to calculate the Bragg angle θ for multiple diffraction number
      function bragg_angles(wavelength::Float64, d::Float64, max::Int)
          angles = Float64[]
          orders = 1:max

          for n in orders
              try
                  angle = bragg_angle(wavelength, d, n)
                  push!(angles, angle)
              catch e
                  @warn "Error calculating Bragg angle for order $n: $e"
              end
          end

          return orders, angles
      end
```

Programme 12.4 Definition the variables and the functions

```
function bragg_angle(wavelength::Float64, d::Float64, n::Int)
    # Bragg's law: n * λ = 2 * d * sin(θ)
    # Rearranged to: sin(θ) = (n * λ) / (2 * d)

    sin_theta = (n * wavelength) / (2 * d)

    if sin_theta > 1.0 || sin_theta < -1.0
        error("Invalid value for sin(θ), cannot calculate angle.")
    end

    theta_rad = asin(sin_theta)

    theta_deg = rad2deg(theta_rad)

    return theta_deg
end

orders, angles = bragg_angles(wavelength, d, max)

for (n, angle) in zip(orders, angles)
    @printf("For n=%d, Bragg angle θ = %.3f°\n", n, angle)
end

Interplanar spacing (in meters):
stdin>  2e-10
Maximum diffraction number :
stdin>  8
For n=1, Bragg angle θ = 2.206°
For n=2, Bragg angle θ = 4.416°
For n=3, Bragg angle θ = 6.632°
For n=4, Bragg angle θ = 8.859°
For n=5, Bragg angle θ = 11.099°
For n=6, Bragg angle θ = 13.356°
For n=7, Bragg angle θ = 15.635°
For n=8, Bragg angle θ = 17.939°
```

Programme 12.5 Output for Bragg Law

$$m\left(\frac{d}{dt}+\frac{1}{\tau}\right)\vec{v} = -e(\vec{E}+\frac{1}{c}(\vec{v}\times\vec{B})) \tag{12.7}$$

If we take a static magnetic field in the z direction and when a steady state in the electric field occurs, we obtain the drift velocities given by Eq. 12.8, where w_c is the cyclotron frequency.

$$v_x = -\frac{e\tau}{m}E_x - w_c\tau v_y$$

$$v_y = -\frac{e\tau}{m}E_y + w_c\tau v_x$$

$$v_z = -\frac{e\tau}{m}E_z \tag{12.8}$$

The Hall Effect, is the electric field generated between the two surfaces in the $\vec{j}\times\vec{B}$ direction of a conductor through which a current $\vec{j}$ passes in a magnetic field $\vec{B}$.

Let us consider a bar conductor placed in an electric field E_x in the x direction and a magnetic field B_z in the z direction. Since the current cannot leave the bar in the y direction, the velocity change in the y direction must be zero.

If Eq. 11.8 is organised in this way, the Eq. 12.8 returns to Eq. 12.9.

$$E_y = -w_c\tau E_x = -\frac{eB\tau}{mc}E_x \tag{12.9}$$

Here let us define a magnitude for the Hall coefficient as in Eq. 12.8.

$$R_H = \frac{E_y}{j_x B} \tag{12.10}$$

According to Ohm's law, using $j_x = ne^2\tau E_x/m$, the Hall coefficient is obtained as in the Eq. 12.10 [1, 2, 5] (Programme 12.6).

$$R_H = -\frac{\frac{eB\tau E_x}{mc}}{\frac{ne^2\tau E_x B}{m}} = -\frac{1}{nec} \tag{12.11}$$

12.5 Madelung Energy

The Madelung energy is the electrostatic potential energy in an ionic crystal due to the interaction between all of its charged particles (ions). This energy arises from the long-range Coulomb interactions between the positive and negative ions in the

```
[ ]: #Hall Effect

[41]: using LinearAlgebra

[185… #Hall Effect
# Determination of variables
q = 1.6E-19  # Charge of an electron (in Coulombs)
B = 0.5      # Magnetic field strength (in Tesla)
i = 2.0      # Current (in Amperes)
n = 1E30     # Carrier concentration (m^-3)
L = 0.1      # Length of the conductor (in meters)
W = 0.01     # Width of the conductor (in meters)

function hall_e(i, B, n, q, L)
    # Calculation of Hall coefficient
    Rh = 1 / (n * q)

    # Calculation of Hall voltage
    Vh = Rh * (i * B) / L

    return Rh, Vh
end

function display_results(Rh, Vh)
    println("Hall Coefficient (R_h): ", Rh, " m^3/C")
    println("Hall Voltage (V_h): ", Vh, " Volts")
end

display_results(Rh, Vh)

Hall Coefficient (R_h): 6.25e-12 m^3/C
Hall Voltage (V_h): 6.25e-11 Volts
```

Programme 12.6 Definition of variables and output for hall effect

crystal lattice. It is important in understanding the stability and properties of ionic crystals, such as sodium chloride (NaCl).

For an ionic crystal structure with alternating positive and negative ions, the Madelung constant (M) quantifies the contribution of these Coulomb interactions.

The long-range interaction between two ions with $\pm q$ charges is the $\pm q^2/r$ electrostatic interaction, which is attractive for charges of opposite sign and repulsive for charges of the same sign. In a crystal structure, ions are placed in positions that are compatible with the repulsive force that occurs at short distances between ion nuclei, making the attractive interaction the largest. Although the Van der Waals part of the attractive interaction in ionic crystals makes a very small contribution to the binding energy, the main contribution comes from the electrostatic part and this energy contribution is called Madelung energy.

If the interaction energy between atoms i and j is U_{ij}, a sum term U_i is defined which gives the interaction of the i ion.

$$U_i = \sum_j U_{ij} \tag{12.12}$$

Suppose that U_{ij} can be written in the sum of repulsive potential $\lambda exp(-{}^{r}/_{\rho})$ and a Coulomb potential $\pm q^2/r$ as in Eq. 11.13, where the +sign is for charges of the same sign and the −sign is for opposite charges.

$$U_{ij} = \lambda \cdot exp\left(-\frac{r_{ij}}{\rho}\right) \pm q^2/r_{ij} \tag{12.13}$$

where R is the nearest neighbour distance in the crystal, we can define the p_{ij} ratio as a form $r_{ij} = p_{ij}R$, and then if repulsive interaction is taken into account only between nearest neighbours Eq. 12.13 returns to Eq. 12.14.

$$U_{ij} = \begin{cases} \lambda . e^{\frac{-R}{\rho}} - \frac{q^2}{R} \\ \pm \frac{1}{p_{ij}} \frac{q^2}{R} \end{cases} \tag{12.14}$$

By using Eq. 11.14, we can obtain total energy as shown in Eq. 11.15, where z is the number of nearest neighbours of an ion and α is defined as the Madelung's constant.

$$U_{\text{tot}} = NU_i = N(z\lambda e^{-\frac{R}{\rho}} - \frac{\alpha q^2}{R}) \tag{12.15}$$

$$\alpha \equiv \sum_j \frac{\pm}{p_{ij}} \equiv \text{Madelung costant} \tag{12.16}$$

If we take $\frac{dU_{tot}}{dR} = 0$ at the equilibrium distance, Eq. 12.15 will be obtained.

$$N\frac{dU_i}{dR} = -\frac{Nz\lambda}{\rho} e^{-\frac{R}{\rho}} + \frac{N\alpha q^2}{R^2} = 0 \tag{12.17}$$

If Eq. 12.15 is regularised, Eq. 12.17 will be obtained. This expression will allow us to find the equilibrium distance R_0 if the ρ, λ parameters of the repulsive interaction are known.

$$R_0^2 e^{-\frac{R_0}{\rho}} = \frac{\rho \alpha q^2}{z\lambda} \tag{12.18}$$

Using Eqs. 12.15 and 12.18, the total lattice expression for a crystal with 2N ions at equilibrium distance is given in Eq. 12.19.

$$U_{tot} = -\frac{N\alpha q^2}{R_0}(1 - \frac{\rho}{R_0}) \tag{12.19}$$

The $-\frac{N\alpha q^2}{R_0}$ term in this expression gives the Madelung energy [1, 2, 5].

Programme 12.7 defines functions of Madelung Energy by using functions [3, 4].

In this section, we have tried to address the basic problems and topics in solid state physics. In the problems we have taken, we first preferred to make the physical definition and explanation of the problem and then we presented the programme to you. We aimed to be the basis for writing code for many topics in solid state physics.

```
[29]: using LinearAlgebra

      # Example values for CsCl structure
      M = 1.76267    # Madelung constant for CsCl
      q = 1.602e-19  # Charge of each ion in Coulombs
      r0 = 4.123e-10  # Lattice constant for CsCl in meters

      # Function to calculate Madelung energy for CsCl structure
      function madelung_energy(M::Float64, q::Float64, r0::Float64)

          E_M = -M * (q^2 / r0)
          return E_M
      end

      # Calculation of Madelung energy
      E_M = madelung_energy(M, q, r0)

      # Convert the energy from Joules to eV (1 Joule = 6.242e+18 eV)
      E_M_eV = E_M * 6.242e+18

      # Output the result
      println("The Madelung energy of the CsCl lattice is: ", E_M_eV, " eV")

      The Madelung energy of the CsCl lattice is: -6.848673558098608e-10 eV
```

Programme 12.7 Madelung energy

References

1. Elementary Solid State Physics, M. Ali OMAR, 1993, ISBN:0-201-60733-6. Addison-Wesley Publishing Company, Inc. Hall Effect chapter 4 and 5; Madelung Energy Chapter 1; Bragg Law chapter 2 ; Wigner-Seitz Unit Cell chapter 5
2. C. Kittel, Introduction to Solid State Physics, 6th edition. 1986, Jhon Wiley & Sons, Inc. Hall Effect chapter 6 ;Madelung Energy Chapter 3; Bragg Law chapter 2 ; Wigner- Seitz Unit Cell chapter 1 and 2
3. Julia-Manual-Functions, https://docs.julialang.org/en/v1/manual/functions/
4. Julia-Developer Documentation-Decumentation of Julia's Internals. https://docs.julialang.org/en/v1/devdocs/functions/
5. Foundamentals of Physics (Extended Edition), David Halliday and Robert Reisnick, 10th Edition. ISBN 978-1-11823072-5. 2007. Hall Effect Chapter 28

Chapter 13
Julia Programming Applications in High Energy Physics

13.1 Introduction

Particle Physics or High Energy Physics (HEP) oriented Julia packages can, via UnROOT.jl, already read HEP's major file formats, including TTree and RNTuple formats. Interfaces to some of HEP's major software packages, such as through Geant4.jl, are available too. Jet reconstruction algorithms in Julia show excellent performance. Several full HEP analyses have been performed in Julia.

In particle physics, quantum electrodynamics (QED) is the relativistic quantum field theory of electrodynamics. It describes how light and matter interact and is the first theory where full agreement between quantum mechanics and special relativity is achieved.

13.2 Properties of the Higgs Boson

Corpuscles.jl is a package which gives easy access to particle properties and identification codes summarised and defined by the Particle Data Group (PDG) collaboration. The cleaned CSV versions of these data are provided by courtesy of the Scikit-HEP project and are part of the Particle Python module which inspired us to create a similar package for the Julia Language. Corpuscles.jl is by far not as feature complete as ***Particle***, but we add functionality continuously, as needed.

The Particle structure can be used to create a particle. If an integer value is passed, it will be interpreted as PDG ID [1], which is the primary particle encoding in Corpuscles.jl [2].

In our current description of Nature, every particle is a wave in a field. The most familiar example of this is light: light is simultaneously a wave in the electromagnetic field and a stream of particles called photons. In the Higgs boson's case, the field came first. The Higgs field was proposed in 1964 as a new kind of field that fills

R. G. Türeci et al., *Julia Programming for Physics Applications*,
https://doi.org/10.1007/978-3-031-84716-5_13

```
[7]: using Corpuscles

     p = Particle(25)
     print(p)
     typeof(p.mass)

     Name:    H
     PDG ID:  25
     LaTeX:   $H^{0}$
     Status:  Common
     Width = -1.0 MeV ± -1.0 MeV
     Q (charge) = 0//1 e
     Mass = 125100.0 MeV ± 140.0 MeV
[7]: Corpuscles.MeasuredValue{L² M T⁻²}
```

Programme 13.1 The code of width, mass, and charge of Higgs boson

Table 13.1 The ID number of quarks from PDG

Quarks	
d	1
u	2
s	3
c	4
b	5
t	6

the entire Universe and gives mass to all elementary particles. The Higgs boson is a wave in that field. Its discovery confirms the existence of the Higgs field.

In Programme 13.1, we want to find the properties of the Higgs boson such as width, mass, and charge using Corpuscles.jl [3].

13.3 Finding the Properties of U and D Quarks

Table 13.1 shows that the ID number of quarks from PDG. Firstly, we can find the properties of width, charge, mass d and u quarks in below Programme 13.2 [2].

13.4 Production of Four Momentum in Particle Physics

In particle physics, quantum electrodynamics (QED) is the relativistic quantum field theory of electrodynamics. It describes how light and matter interact and is the first theory where full agreement between quantum mechanics and special relativity is achieved.

QuantumElectrodynamics.jl includes the combination of the following two fundamental packages [4].

```
[6]: using Corpuscles
     p = Particle(1)
     print(p)
     p = Particle(2)
     print(p)

     Name:    d
     PDG ID:  1
     LaTeX:   $d$
     Status:  Common
     Width = -1.0 MeV ± -1.0 MeV
     Q (charge) = -1//3 e
     Composition = d
     Isospin = 1//2
     Mass = 4.67 MeV + 0.2 MeV - 0.5 MeV
     P (space parity) = 1
     Name:    u
     PDG ID:  2
     LaTeX:   $u$
     Status:  Common
     Width = -1.0 MeV ± -1.0 MeV
     Q (charge) = 2//3 e
     Composition = u
     Isospin = 1//2
     Mass = 2.16 MeV + 0.3 MeV - 0.5 MeV
     P (space parity) = 1
```

Programme 13.2 The code of properties of quarks such as width, mass, charge, isospin, and parity

- ***QEDbase.jl***: Interfaces and some functionality on abstract types. This package provides the general data structures for calculations in relativistic particle physics [4].

  Current features of ***QuantumElectrodynamics.jl***:

- Generic interface for Lorentz vectors
- concrete implementations of static and mutable Lorentz-vector/four-momentum types
- general Dirac bi-spinors, its adjoint counterpart as well as Dirac matrices
- particle spinors, i.e. solutions of Dirac's equation in momentum space
- Dirac's gamma matrices
- ***QEDcore.jl***: Implementation of core functionality that is needed across all or most content packages [5].

  The content packages:

  QEDprocesses.jl: Scattering process definitions, models, and calculation of cross-sections and probabilities.
  QEDevents.jl: Monte-Carlo event generation for scattering processes.

```
[1]: using Pkg
     using QEDbase
     Pkg.test("QEDbase")
```

```
Test Summary:  | Pass  Total  Time
cross sections |   56     56  4.9s
  4.921885 seconds (7.74 M allocations: 535.284 MiB, 9.86% gc time, 96.87% compilation time)
     Testing QEDbase tests passed
```

Programme 13.3 The test of QEDbase

```
[49]: import Pkg; Pkg.add("QEDbase")
      import Pkg; Pkg.add("QEDcore")
      using QEDbase;
      using QEDcore;

      mass=rand()*10
      px,py,pz=rand(3)
      E=sqrt(px^2+py^2+pz^2+mass^2)
      P=SFourMomentum(E,px,py,pz)

         Resolving package versions...
        No Changes to `~/Project.toml`
        No Changes to `~/Manifest.toml`
      Precompiling project...
        ? Corpuscles
         Resolving package versions...
        No Changes to `~/Project.toml`
        No Changes to `~/Manifest.toml`
      Precompiling project...
        ? Corpuscles

[49]: 4-element SFourMomentum with indices SOneTo(4):
       3.5505772387124845
       0.7950669187297038
       0.15054400475065455
       0.47445776549984486
```

Programme 13.4 The code of production four momentum

QEDfields.jl: Description of classical electromagnetic fields used in background-field approximations.

Initially, we can start to test using Programme 13.3.

If we see the last line as "***Testing QEDbase tests passed***", we can assume that ***QEDbase.jl*** works properly for us [6].

Now we can produce four momentum in particle physics using Programme 13.4.

13.5 Lorentz Four Vector

Momentum is a vector in three dimensions and four momentum is a four vector in spacetime. The contravariant four momentum of a particle with relativistic energy E and three-momentum $\mathbf{p} = (p_x, p_y, p_z) = \gamma m \mathbf{v}$, where $\mathbf{v}$ is the particle's three-velocity and γ the Lorentz factor, is [7]

```
[51]: #Given px, py, pz, m returns lorentz vector e, px, py, pz
      using LorentzVectors
      function fromPxPyPzM(px, py, pz, m)
          e = sqrt(muladd(px, px, py^2) + muladd(pz, pz, m^2))
          return LorentzVector(e, px, py, pz)
      end
      fromPxPyPzM(1,1,1,1)

[51]: LorentzVector{Float64}(2.0, 1.0, 1.0, 1.0)
```

Programme 13.5 The code of Lorentz four vector

```
[53]: # cos theta from Lorentz Vector
      using LorentzVectors
      Base.zero(lv::T) where T<:LorentzVector = T(0,0,0,0)
      mag(lv::LorentzVector) = sqrt(muladd(lv.x, lv.x, lv.y^2) + lv.z^2)
      @inline function CosTheta(lv::LorentzVector)
          fZ = lv.z
          ptot = mag(lv)
          return ifelse(ptot == 0.0, 1.0, fZ / ptot)
      end
      p1 = Vec4(1, 1, 1, 1)
      CosTheta(p1)

[53]: 0.5773502691896258
```

Programme 13.6 The code of conversion from vector to theta angle

$$p = (p^0, p^1, p^2, p^3) = (\frac{E}{c}, p_x, p_{y,}p_z) \tag{13.1}$$

The quantity $m\mathbf{v}$ of above is the nonrelativistic momentum of the particle and m its rest mass. The four-momentum is useful in relativistic calculations because it is a Lorentz covariant vector.

Given the components of a vector and mass of the particle, this code in Programme 13.5 transforms it to a Lorentz four vector [8].

13.6 Changing a Given Four Vector to Cos(θ)

This code returns cos(θ) from a given four vector (p^0, p^1, p^2, p^3) in Programme 13.6. Here θ is the angle from z-axes.

References

1. https://pdg.lbl.gov/2019/reviews/rpp2019-rev-monte-carlo-numbering.pdf
2. https://github.com/JuliaPhysics/Corpuscles.jl
3. https://home.cern/science/physics/higgs-boson
4. https://github.com/QEDjl-project/QEDbase.jl?tab=readme-ov-file
5. https://qedjl-project.github.io/QuantumElectrodynamics.jl/dev/

6. https://juliapackages.com/p/qedbase
7. https://en.wikipedia.org/wiki/Four-momentum
8. https://github.com/JLTastet/LorentzVectors.jl

Chapter 14
Julia Programming Applications in Astrophysics

14.1 Introduction

In this section, we will focus on the basic problems from simple to difficult from a pedagogical point of view related to the subject called waves in physics and the analysis of these problems using the Julia programming language. After some applications, more detailed parts of the analysis will be left as questions for the reader. Thus, it is thought that both the Julia programming language and, more importantly, the physics problem will be discussed in a more understandable way.

14.2 Kepler's Law

Kepler's laws describe the motion of planets in the solar system and can be implemented in Julia for simulation and visualisation. There are three laws:

1. ***Kepler's First Law (Law of Ellipses)***

This law revolutionised our understanding of planetary motion, as it contradicted the long-held belief that planets moved in perfect circles. An ellipse is an elongated circle, so the distance between a planet and the Sun varies as the planet moves along its orbit.

2. ***Kepler's Second Law (Law of Equal Areas)***

This means that a planet moves faster when it is closer to the Sun and slower when it is farther away. The area swept by the line between the planet and the Sun (called the radius vector) is always constant over a given period of time, even though the speed of the planet changes.

R. G. Türeci et al., *Julia Programming for Physics Applications*,
https://doi.org/10.1007/978-3-031-84716-5_14

3. ***Kepler's Third Law (Law of Harmonies)***

This law establishes a relationship between the time it takes a planet to complete an orbit and the size of its orbit. The farther a planet is from the Sun, the longer it takes to complete an orbit.

Consider a planet of mass *Mp* moving around the Sun of mass M_S in a circular orbit. Because of the gravitational force exerted by the Sun on the planet is a radially directed force that keeps the planet moving in a circle, we can apply Newton's second law ($F =$ ma) to the planet as shown in Eq. 14.1.

$$G\frac{M_S M_P}{r^2} = \frac{M_P v^2}{r} \tag{14.1}$$

Because the orbital speed v of the planet is simply $2\pi\ r/T$, where T is its period of revolution, the Eq. 14.1 becomes [1, 2]

$$G\frac{M_S}{r^2} = \frac{\left(\frac{2\pi r}{T}\right)^2}{r} \tag{14.1}$$

$$T^2 = \frac{4\pi^2}{GM_S}r^3 \tag{14.2}$$

where in Eq. 14.2, K is a constant given by in Eq. 14.3.

$$K = \frac{4\pi^2}{GM_S} = \frac{T^2}{r^3} = 2{,}97 \times 10^{-19}\,\frac{\mathrm{s}^2}{\mathrm{m}^3} \tag{14.3}$$

Programme 14.1 shows Kepler's third law by using functions [3, 4].

14.3 Stefan-Boltzmann Law

The Stefan-Boltzmann Law is a fundamental principle in physics that describes the power radiated by a perfect black body in terms of its temperature. It states that the total energy radiated per unit surface area of a black body is directly proportional to the fourth power of its absolute temperature T.

The Stefan-Boltzmann Law is mathematically expressed as:

$$E = \sigma A T^4 \tag{14.4}$$

where in Eq. 14.4 E is the radiated energy per unit area also as known as the radiant emmitance (*in* $\frac{\mathrm{W}}{\mathrm{m}^2}$), σ is the Stefan-Boltzmann constant, σ = 5.670374419 × 10 − 8 W m^{-2} K^{-4}, A is the surface area of the radiating body (*in square meters*, m^2), T is the absolute temperature of the body (*in kelvins, K*) [5].

Programme 14.2 shows the calculation of Stephan-Bolltzmann's Law.

```
[11]: using LinearAlgebra

# Determination of Variables
G = 6.67430e-11 # Gravitational constant in m^3/kg/s^2
M = 1.989e30     # Mass of the Sun in kg
AU = 1.496e11    # 1 AU in meters (Astronomical Unit)

# Semi-major axis for Saturn (in meters)
a_sat = 9.58 * AU

# The orbital period (T) from Kepler's Third Law
function orbital_period(a::Float64)
    T_squared = (4 * π^2 * a^3) / (G * M)
    T = sqrt(T_squared)
    return T  # Orbital period in seconds
end

function seconds_to_years(seconds::Float64)
    return seconds / (60 * 60 * 24 * 365)
end

# The orbital period of Saturn
T_sat_seconds = orbital_period(a_sat)  #  (in seconds)
T_sat_years = seconds_to_years(T_sat_seconds)  # Convert to years

println("Orbital period of Saturn: ", T_sat_years, " years")

Orbital period of Saturn: 29.66870989100826 years
```

Programme 14.1 Kepler's third law

```
3]: # Constants
σ = 5.670374419e-8  # Stefan-Boltzmann constant in W/m^2/K^4
r_earth = 6.371e6     # Radius of the Earth in meters
T_earth = 288         # Average surface temperature of the Earth in Kelvin

# Surface area of the Earth (A = 4πr^2)
A_earth = 4 * π * r_earth^2

# Total power radiated (P = σ * A * T^4)
P_earth = σ * A_earth * T_earth^4

# Print the result in watts
println("Total power radiated by Earth: ", P_earth, " watts")

Total power radiated by Earth: 1.9897877913061648e17 watts
```

Programme 14.2 Stephan-Boltzmann's Law

14.4 Hubble's Law

Hubble's Law is a key observation in cosmology that describes the expansion of the universe. It states that the farther a galaxy is from us, the faster it is moving away from us. This relationship between a galaxy's distance and its velocity is linear, which means that as the distance increases, the velocity of recession increases proportionally.

The Hubble constant quantifies the rate of expansion of the universe, but its exact value is still a subject of research, with different methods giving slightly different

```
[3]: using LinearAlgebra

     hubble_constant = 70.0  # Hubble constant

     println("Distance (in megaparsec): ")
     dist= parse(Float64, readline())
     function hubble_vel(distance::Float64, hubble_constant::Float64)
         return hubble_constant * dist
     end

     # Calculation of the recession velocity
     velocity = hubble_vel(dist, hubble_constant)

     # Print the result
     println("The recession velocity of the galaxy is ", velocity, " km/s.")

     Distance (in megaparsec):
     stdin> 200
     The recession velocity of the galaxy is 14000.0 km/s.
```

Programme 14.3 Hubble's Law

results, the second importance of Hubble's law is; Hubble's law supports the idea that the universe is expanding.

Hubble used observations of distant galaxies and their redshifts (shifts in the wavelength of light toward the red end of the spectrum, which indicates that an object is moving away) to conclude that galaxies farther from Earth had greater redshifts, meaning they were receding faster. This observation supported the idea that the universe was expanding, a conclusion that would later be confirmed by other observations, such as the discovery of cosmic microwave background radiation.

Hubble's discovery was key to the development of the Big Bang theory, which posits that the universe began and has been expanding ever since.

This law describes the relationship between the distance to a galaxy and its recession velocity as shown in Eq. 14.5, where v is the velocity at which the galaxy is receding (in km/s), d is the distance to the galaxy (in *megaparsecs*, Mpc) and H_0 is the Hubble constant, which describes the rate of expansion of the universe [6]

$$v = H_0 d \tag{14.5}$$

In Programme 14.3, the calculations are made with using the function method [4].

14.5 Luminosity of a Star

Luminosity refers to the total amount of energy a star or other astronomical object emits per unit of time. It is a measure of the intrinsic brightness of an object, independent of its distance from the observer. For stars, luminosity is often measured in terms of the Sun's luminosity (L_s), which is approximately 3.828 × 1026 W (Programme 14.4).

```
[11]: using LinearAlgebra

      # Define constants and variables
      σ = 5.67e-8  # Stefan-Boltzmann constant in W/m^2/K^4

      println("Radius of the star (in meters): ")
      R= parse(Float64, readline())
      println("Temperature (in Kelvin): ")
      T= parse(Float64, readline())

      # Calculate the luminosity
      L = 4 * π * R^2 * σ * T^4


      println("The luminosity of the star is ", L, " watts.")

      Radius of the sun (in meters):
      stdin>  10e8
      Temperature (in Kelvin):
      stdin>  5000
      The luminosity of the star is 4.4532075864635316e26 watts.
```

Programme 14.4 Luminosity of the star

The luminosity L of a star can be calculated using the Stefan-Boltzmann law, which relates a star's luminosity to its temperature and radius as shown in Eq. 14.6. where; L, luminosity of the star (*in Watts*), R, radius of the star (*in meters*), σ, Stefan-Boltzmann constant and T, temperature of the star's surface (*in Kelvin*).

$$L = 4\pi R^2 \sigma T^4 \tag{14.5}$$

In this chapter, we have dealt with problems in astronomy and astrophysics. We have explained the physics concepts that are mostly mentioned in our daily life without going too deep, and we have mentioned Julia programme codes related to these topics. The examples can be varied a lot, and readers will have the opportunity to make these codes more comprehensive.

References

1. Physics For Scientists and Engineers with Modern Physics, 5th Edition, Raymond A. Serway, Robert J. Beichner, Saunders College Publishing, 2000. Kepler's Law, Chapter 13.
2. Foundamentals of Physics (Extended Edition), David Halliday and Robert Reisnick, 10th Edition. ISBN 978-1-11823072-5. 2007. Kepler's Law, Chapter 13
3. Julia-Manual-Functions, https://docs.julialang.org/en/v1/manual/functions/
4. Julia-Developper Documentation-Documentation of Julia's Internals-Julia Functions, https://docs.julialang.org/en/v1/devdocs/functions/
5. Foundamentals of Physics (Extended Edition), David Halliday and Robert Reisnick, 10th Edition. ISBN 978-1-11823072–5. 2007. Stefan-Boltzmann Constant, Chapter 18
6. Foundamentals of Physics (Extended Edition), David Halliday and Robert Reisnick, 10th Edition. ISBN 978-1-11823072-5. 2007. Hubble's Law, Chapter 44

Chapter 15
Julia Programming Applications in Statistics

15.1 Introduction

15.1.1 Monte-Carlo Simulation and Random Numbers

Generally many of the codes make use of pseudo-random number generation, which is coupled with the so-called Monte Carlo simulation method for obtaining numerical estimates. The Monte Carlo (MC) method is a numerical technique for solving statistical problems and involves the use of random numbers to simulate, for instance, measurements in a specific geometry. The phrase Monte Carlo comes from the European province in Monaco where gambling is popular and refers to the random nature of such activities. Computerised random number generators produce what are called pseudo-random numbers, in that the numbers are produced by a deterministic process that gives exactly the same sequence of numbers every time.

The simplest use is just calling *rand()*, which returns a random in the interval [0, 1], when used without input arguments. The random numbers that we generate using by Julia, also most random numbers used in any other scientific computing platform, are pseudorandom which are not really random but rather appear random, Klok and Nazarathy [1].

15.2 Pseudorandom Number Generation

We can set a random seed by using the *Random.seed! ()* in Programme 15.1 [1].

In this example, all settings of the seed are completely different sequence. In the new version of Julia, the syntax to set a seed to the random number generator changes. You are now supposed to use seed!() from the Random package. However, keep in mind that you must use it every time you are using the random number generator.

R. G. Türeci et al., *Julia Programming for Physics Applications*,
https://doi.org/10.1007/978-3-031-84716-5_15

```
[6]: using Random

     Random.seed!(74)
     println("Seed 74: ",rand(),"\t", rand(), "\t", rand())
     Random.seed!(75)
     println("Seed 75: ",rand(),"\t", rand(), "\t", rand())
     Random.seed!(74)
     println("Seed 74: ",rand(),"\t", rand(), "\t", rand())

     Seed 74: 0.6609753391503894     0.8170202503578953      0.9329724337150506
     Seed 75: 0.2423950634469636     0.38163725329739184     0.14188003713956232
     Seed 74: 0.6609753391503894     0.8170202503578953      0.9329724337150506
```

Programme 15.1 The code of pseudorandom number generation

15.3 Independent Events

Suppose that we flip two fair coins simultaneously on opposite sides of a room. Intuitively, the way one-coin lands does not affect the way the other coin lands. The mathematical concept that captures this intuition is called independence. Events A and B are independent if and only if [2]:

$$P\,(A \cap B) = P(A) \cdot P(B) \tag{15.1}$$

Generally, independence is something you assume in modelling a phenomenon or wish you could realistically assume. We can see independent events example using random number in Programme 15.2, Klok and Nazarathy [1].

We get the following results:

$$\begin{aligned} P(13) &= 0.062486,\ P(1\_) = 0.625017, \\ P(\_3) &= 0.125023, \end{aligned}$$

```
[34]: using Random
      Random.seed!(1)
      numbers = 10:25
      N = 10^6


      firstDigit(x) = Int(floor(x/10))
      secondDigit(x) = x%10
      numThirteen, numFirstIsOne, numSecondIsThree = 0, 0, 0

      for _ in 1:N
          X = rand(numbers)
          global numThirteen += X == 13
          global numFirstIsOne += firstDigit(X) == 1
          global numSecondIsThree += secondDigit(X) == 3
      end
      probThirteen, probFirstIsOne, probSecondIsThree = (numThirteen, numFirstIsOne, numSecondIsThree)./N
      #println("P(13)=",probThirteen)
      println("P(13) = ", (probThirteen),
          "\nP(1_) = ",(probFirstIsOne),
          "\nP(_3) = ", (probSecondIsThree),
          "\nP(1_)*P(_3) = ",(probFirstIsOne*probSecondIsThree),
           "\nP(_3) = ", (probSecondIsThree),
           "\nP(1_)*P(_3) = ",(probFirstIsOne*probSecondIsThree))
```

Programme 15.2 The code of independent events using random number

$$P(1\_) * P(\_3) = 0.078141500391,\ P(\_3) = 0.125023,$$
$$P(1\_) * P(\_3) = 0.07814150039.$$

In this code[1], line 7 defines a function that returns the first digit of our number through the use of the floor() function, and converts the resulting value to an integer type. Line 8 defines a function that uses the modulus operator % to return the second digit of our number. In line 10 we initialise three placeholder variables, which represent the number chosen, and its first and second digits, respectively. Lines 12–17 contain the core logic of this example, where N random digits are generated. For each random digit, X that is generated, lines 14, 15, and 16 increment the count by 1 if the specified condition is met. Lines 19–20 evaluate the total proportions.

15.4 Estimating π-Number with Monte Carlo

We want to estimate the value of π in Programme 15.3, Klok and Nazarathy [1]. We think that the area of one-quarter section of the unit circle is π/4. We can calculate the proportion of total points that fall within the quarter circle, if we generate random some points, (x, y) in the unit box, [0, 1] × [0, 1].

The seed of the random number generator is set with Random.seed!(). This is done to ensure that each time the code is run the estimate obtained is the same. In Line 4, the number of repetitions, N, is set. Line 5 generates an array of arrays. That is, the pair, [rand(), rand()] is an array of random coordinates in [0, 1] × [0, 1]. Line 6 filters those points to use for the numerator of π. Line 7 creates the analogous out data array. It is not used for the estimation but is used in plotting. Line 8 calculates the approximation, with length() used for the numerator of π and N for the denominator. Lines 11–13 are used to make Fig. 15.3.

```
Julia: A fresh approach to technical computing.

[1]: using Random, LinearAlgebra, Plots; pyplot()
     Random.seed!()
     N=10^6
     data = [[rand(),rand()] for _ in 1:N]
     inindata = filter((x)-> (norm(x) <= 1), data)
     outoutdata = filter((x)-> (norm(x) > 1), data)
     piApprox = 4*length(inindata)/N
     println("Pi Estimate: ", piApprox)
     scatter(first.(inindata),last.(inindata), c=:blue, ms=1, msw=0)
     scatter!(first.(outoutdata),last.(outoutdata), c=:red, ms=1, msw=0,
     xlims=(0,1), ylims=(0,1), legend=:none, ratio=:equal)

     Pi Estimate: 3.139604
```

Programme 15.3 Estimation of π-number with Monte Carlo

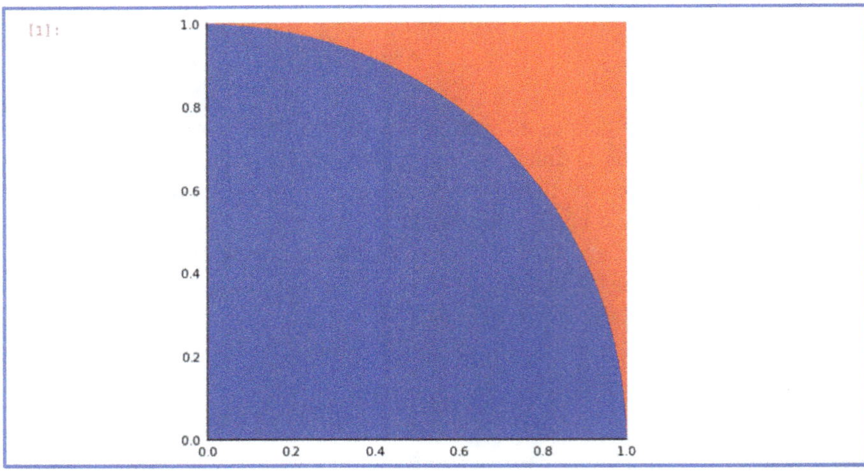

Fig. 15.3 Estimating π-number with Monte Carlo

15.5 Types of Random Variables

In probability theory and statistics, a **probability distribution** is the mathematical function that gives the probabilities of occurrence of possible **outcomes** for an experiment.

We consider a sample space Ω which consists of six names. We assume that the probability function **P**(·), assigns uniform probabilities to each of the names. Let now, we define random variable as X: Ω → Z, be the function that counts the number of letters in name. The question is then finding:

$$p(x) := \mathbf{P}(\mathbf{X} = \mathbf{x})\text{for}x \in Z \tag{15.2}$$

The function $p(x)$ represents the probability distribution of the random variable X. Since x measures name lengths, X is a discrete random variable, and its probability distribution may be represented by a Probability Mass Function (PMF), such as $p(x)$.

We can carry out a simulation of many such random experiments, yielding many replications of the random variable X, which we then use to estimate $p(x)$ in Programme 15.4, Klok and Nazarathy [1] (Fig. 15.4).

In line 3, we created the array names containing different character lengths. Note that two names have five characters, namely, "Hamdi" and "Ilkay", while there is no name for seven characters. We define the function ***randomName()*** which randomly selects in line 4, with equal probability, an element from the array names. In line 5, we specify that we will count names of three to eight characters in length. Line 6 specifies how many random experiments of choosing a name we will perform. Line 7 uses a comprehension and the function ***length()*** to count the length of each random name, and stores the results in the array *sampleLengths*. Here the Julia function

```
[65]: using StatsBase, Plots; pyplot()

      names = ["Hamdi","Ali","Elif","Ilkay","Gokhan","JuliaHEP","Ankara"]
      randomName() = rand(names)
      X = 3:8
      N = 10^6
      sampleLengths = [length(randomName()) for _ in 1:N]

      bar(X,counts(sampleLengths)/N, ylims=(0,0.35),
      xlabel="The name length", ylabel="The estimated p(x)", legend=:none)
```

Programme 15.4 The discreate probability distribution using {3, 4, 5, 6, 8} values

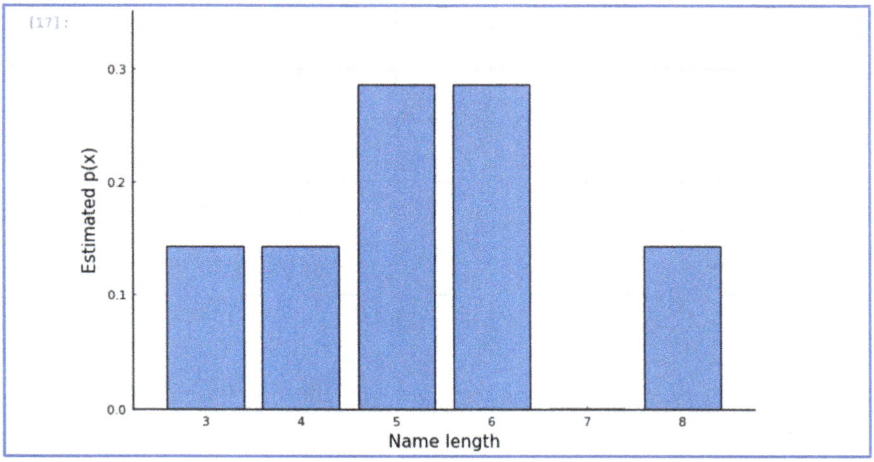

Fig. 15.4 The discrete probability distribution using {3, 4, 5, 6, 8} values

length() is the analogy of the random variable. That is, it is a function of the sample space, Ω, yielding a numerical value. Line 9 uses the function ***counts()*** to count how many words are of length 3, 4, up to 8. The ***bar()*** function is then used to plot a bar chart of the proportion of counts for each word length. Two key observations can be made. If there are no words of length 7 were selected, as no name in our original array had a length of 7.

15.6 Discrete and Continuous Distributions

Random variable is a fundamental concept in statistics that bridges the gap between theoretical probability and real-world data. **Random variable** in statistics is a function that assigns a real value to an outcome in the sample space of a random experiment. For example: if you roll a die, you can assign a number to each possible outcome [3].

There are two basic types of random variables:

- Discrete Random Variables (which take on specific values
- Continuous Random Variables (assume any value within a given range).

The discrete probability distribution appearing on the left in Fig. 15.3 can be represented mathematically by the probability mass function.

$$p(x) = \begin{cases} 0.25 & \text{for } x = 0 \\ 0.25 & \text{for } x = 1 \\ 0.50 & \text{for } x = 2 \end{cases} \tag{15.3}$$

The smooth continuous probability distribution is defined by the probability density function,

$$f_1(x) = \frac{3}{5}\left(1 - x^2\right) = 0 \text{ for } -1 \leq x \leq 1 \tag{15.4}$$

Finally, the triangular probability distribution is defined by the probability density function,

$$f_2(x) = \begin{cases} x + 1 & \text{for } x \in [-1{,}0] \\ 1 - x & \text{for } x \in (0{,}1] \end{cases} \tag{15.5}$$

Note that for both the probability mass function and the probability density function, it is implicitly assumed that $p(x)$ and $f(x)$ are zero for x values not specified in the equation (Programme 15.5 and Fig. 15.5).

It can be verified that for the discrete distribution,

$$\sum_x \boldsymbol{p(x)} = 1 \tag{15.6}$$

and for the continuous distributions,

$$\int_{-\infty}^{\infty} f_i(x)dx = 1 \text{ for } i = 1, 2 \tag{15.7}$$

```
[13]: import Pkg; Pkg.add("Measures")
      using Plots, Measures; pyplot()
      pDiscrete = [0.25, 0.25, 0.50]
      xGridD = 0:2

      pContinuous(x) = 3/5*(1 - x^2)
      xGridC = -1:0.01:1

      pContinuous2(x) = x < 0 ? x+1 : 1-x

      p1 = plot(xGridD, line=:stem, pDiscrete, marker=:circle, c=:red, ms=6, msw=0)
      p2 = plot(xGridC, pContinuous.(xGridC), c=:red)
      p3 = plot(xGridC, pContinuous2.(xGridC), c=:red)

      plot(p1, p2, p3, layout=(1,3), legend=false, ylims=(0,1.1), xlabel="x",
      ylabel=["PROBABILITY" "DENSITY" "DENSITY"], size=(1300, 500), margin=6mm)
```

Programme 15.5 The code of discrete and continuous distributions

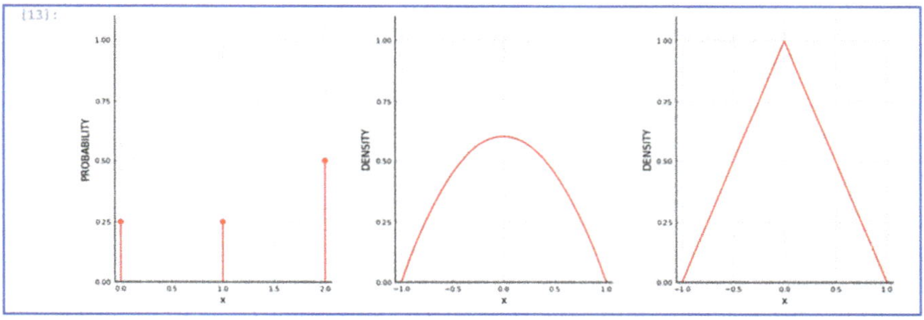

Fig. 15.5 Plotting discrete and continuous distributions

We have defined an array specifying the PMF of our discrete distribution in line 3 and define functions specifying the PDFs of our continuous distributions in lines 6 and 9 in Programme 15.4. In lines 11–16, we create all plots of each of our distributions. Note that we use the *line=:stem* argument together with marker=:circle in the discrete case, Klok and Nazarathy [1].

References

1. Hayden Klok, Yoni Nazarathy, (2021), Statistics with Julia: Fundamentals for Data Science, Machine Learning and Artificial Intelligence
2. http://web.mit.edu/neboat/Public/6.042/independence.pdf
3. https://www.geeksforgeeks.org/random-variable/

GPSR Compliance
The European Union's (EU) General Product Safety Regulation (GPSR) is a set of rules that requires consumer products to be safe and our obligations to ensure this.

If you have any concerns about our products, you can contact us on

ProductSafety@springernature.com

In case Publisher is established outside the EU, the EU authorized representative is:

Springer Nature Customer Service Center GmbH
Europaplatz 3
69115 Heidelberg, Germany

www.ingramcontent.com/pod-product-compliance
Lightning Source LLC
Chambersburg PA
CBHW072035110625
27980CB00025B/42
* 9 7 8 3 0 3 1 8 4 7 1 5 8 *